Lecture Notes in Artificial Intelligence 8875

Subseries of Lecture Notes in Computer Science

LNAI Series Editors

Randy Goebel
University of Alberta, Edmonton, Canada
Yuzuru Tanaka
Hokkaido University, Sapporo, Japan
Wolfgang Wahlster
DFKI and Saarland University, Saarbrücken, Germany

LNAI Founding Series Editor

Joerg Siekmann
DFKI and Saarland University, Saarbrücken, Germany

M. Narasimha Murty   Xiangjian He
Raghavendra Rao Chillarige   Paul Weng (Eds.)

Multi-disciplinary Trends in Artificial Intelligence

8th International Workshop, MIWAI 2014
Bangalore, India, December 8-10, 2014
Proceedings

 Springer

Volume Editors

M. Narasimha Murty
Indian Institute of Science
Department of Computer Science and Automation
Bangalore 560012, India
E-mail: mnm@csa.iisc.ernet.in

Xiangjian He
University of Technology Sydney
School of Computing and Communications
15 Broadway, Ultimo, NSW 2007, Australia
E-mail: xiangjian.he@uts.edu.au

Raghavendra Rao Chillarige
University of Hyderabad
School of Computer and Information Sciences
Hyderabad 500046, India
E-mail: crrcs@uohyd.ernet.in

Paul Weng
University Pierre et Marie Curie, LIP6
4 Place Jussieu, 75005 Paris, France
E-mail: paul.weng@lip6.fr

ISSN 0302-9743   e-ISSN 1611-3349
ISBN 978-3-319-13364-5   e-ISBN 978-3-319-13365-2
DOI 10.1007/978-3-319-13365-2
Springer Cham Heidelberg New York Dordrecht London

Library of Congress Control Number: 2014954258

LNCS Sublibrary: SL 7 – Artificial Intelligence

Typesetting: Camera-ready by author, data conversion by Scientific Publishing Services, Chennai, India

Printed on acid-free paper

Springer is part of Springer Science+Business Media (www.springer.com)

Preface

The MIWAI workshop series is an annual workshop that was started by the Mahasarakham University in Thailand in 2007 as the Mahasarakham International Workshop on Artificial Intelligence. It has progressively emerged as an international workshop with participants from around the world. Thus, in 2012, MIWAI was held in Ho Chi Minh City, Vietnam, and MIWAI 2013 was held in Krabi, Thailand.

Continuing its tradition, the MIWAI series of workshops offers a forum where artificial intelligence (AI) researchers and practitioners can discuss cutting-edge research ideas and present innovating research applications. It also strives to elevate the standards of AI research by providing researchers with feedback from an internationally renowned Program Committee.

AI is a broad research areas. Theory, methods, and tools in AI sub-areas such as cognitive science, computational philosophy, computational intelligence, game theory, multi-agent systems, machine learning, natural language, robotics, representation and reasoning, speech and vision, along with their diverse applications in domains including big data, biometrics, bio-informatics, decision-support systems, e-commerce, e-health, e-learning, knowledge management, privacy, recommender systems, security, software engineering, spam-filtering, or telecommunications, and Web services are topics of interest to MIWAI.

This year, MIWAI 2014 reflected this broad coverage as the submissions were wide ranging and covered both theory as well as applications. This volume contains the research papers presented at the 8th Multi-disciplinary International Workshop on Artificial Intelligence (MIWAI) held during December 8–10, 2014, in Bangalore, India.

MIWAI 2014 received 44 full papers from many countries including Algeria, Bangladesh, Canada, Germany, India, Korea, Morocco, Poland, UK, and Vietnam. Following the success of previous MIWAI conferences, MIWAI 2014 continued the tradition of a rigorous review process. All submissions were subject to a brief review by the program and general chairs to ensure a blind and fair review. Every submission was reviewed by at least two Program Committee members and domain experts. Additional reviews were sought when necessary. At the end, a total of 22 papers were accepted with an acceptance rate of 50%. Some of the papers that were excluded from the proceedings showed promise, but had to be rejected to maintain the quality of the proceedings. We would like to thank all authors for their submissions. Without their contribution, this workshop would not have been possible.

We are grateful to Prof. C. A. Murthy for accepting our invitation to deliver the keynote talk. Special thanks to all the invited speakers Prof. P.S. Sastry, Prof. B.S. Daya Sagar, Dr. Biplav Srivastava and the tutorial speakers Prof. P. Krishna Reddy and Dr. Chattrakul Sombattheera. We are indebted to the

Program Committee members and external reviewers for their effort in ensuring a rich scientific program.

We acknowledge the use of the EasyChair Conference System for the paper submission, review, and compilation of the proceedings. We are also thankful to all our sponsors: IBM India Pvt. Ltd, Institute for Development and Research in Banking Technology, Allied Telesis India Pvt. Ltd, Locuz Enterprise Solutions Ltd. Last but not the least our sincere thanks to Alfred Hofmann, Anna Kramer, and the excellent LNCS team at Springer for their support and cooperation in publishing the proceedings as a volume of the *Lecture Notes in Computer Science.*

September 2014   M. Narasimha Murty
                                                      Xiangjian He
                                              C. Raghavendra Rao
                                                       Paul Weng

Organization

Steering Committee

Arun Agarwal   University of Hyderabad, India
Rajkumar Buyya   University of Melbourne, Australia
Patrick Doherty   University of Linkoping, Sweden
Jerome Lang   University of Paris, Dauphine, France
James F. Peters   University of Manitoba, Canada
Srinivasan Ramani   IIIT Bangalore, India
C Raghavendra Rao   University of Hyderabad, India
Leon Van Der Torre   University of Luxembourg, Luxembourg

Conveners

Richard Booth   University of Luxembourg, Luxembourg
Chattrakul Sombattheera   Mahasarakham University, Thailand

General Co-chairs

M. Narasimha Murty   Indian Institute of Science, India
Xiangjian He   University of Technology Sydney, Australia

Program Co-chairs

C Raghavendra Rao   University of Hyderabad, India
Paul Weng   Université Paris 6 (UPMC), France

Program Committee

Arun Agarwal   University of Hyderabad, India
Samir Aknine   Claude Bernard University of Lyon 1, France
Ricardo Aler   Universidad Carlos III de Madrid, Spain
Rafah Mohammed Almuttairi   University of Babylon, Iraq
Grigoris Antoniou   University of Huddersfield, UK
Costin Badica   University of Craiova, Romania
Raj Bhatnagar   University of Cincinnati, USA
Hima Bindu   Vishnu Institute of Technology, India
Laor Boongasame   Bangkok University, Thailand
Veera Boonjing   King Mongkut's Institute of Technology
     Ladkrabang, Thailand
Richard Booth   University of Luxembourg, Luxembourg

Publicity Co-chairs

Rajeev Wankar   University of Hyderabad, India
Manish Joshi   North Maharashtra University, India

Local Organizing Committee

B.L. Muralidhara   Bangalore University, India
Dilip Kumar S. M.   University Visvesvaraya College of Engineering,
                                   India

Web Administrator

Panich Sudkhot   Mahasarakham University, Thailand

Additional Reviewers

Jednipat Moonrinta

Table of Contents

"Potential Interval of Root" of Nonlinear Equation: Labeling Algorithm

Vijaya Lakshmi V. Nadimpalli[1], Rajeev Wankar[2], and Raghavendra Rao Chillarige[2]

[1] ACRHEM, University of Hyderabad, Hyderabad, India – 500046
[2] SCIS, University of Hyderabad, Hyderabad, India – 500046
nvvlakshmi@gmail.com, {wankarcs,crrcs}@uohyd.ernet.in

Abstract. Novel Sequence Generating Algorithm (SGA) and Potential Interval Qualifier Algorithm (PIQA) are designed to classify potential interval estimates of a given nonlinear equation into intervals possessing roots and intervals containing extrema. Using trisection method, SGA is developed to generate conjugate pair of sequences that converge to a point in the interval. Further, PIQA qualifies each interval into interval enclosing a root or interval containing extrema. If the interval contains a root, the multiplicity of root is also obtained. The proposed methodologies have been implemented and demonstrated through a set of benchmark functions to illustrate the effectiveness.

Keywords: Convergence, extrema, Genetic Algorithms, interval based method, interval estimates, root, sequence.

1 Introduction

A number of real world problems in different areas of science and engineering are often reduced to the problem of finding all the real roots and/or the extrema of a function in a given interval. Most of the conventional numerical methods like Newton-Raphson method, secant method have some drawbacks such as sensitivity to initial guess and oscillatory behavior of root etc. Generally interval based root computation methods are known to be robust but they are slow. Finding roots through interval based bracketing root method called Bisection method is discussed in [1] by subdividing the given interval in to 'n' equal intervals. Most of the real world problems are likely to be complex and knowledge about the function is generally not known in advance and hence we do not know how many subdivisions of intervals are needed for a given problem in the region of interest. Various Metaheuristic optimization methods such as Genetic Algorithms (GA) [2, 3], Particle Swarm Optimization (PSO) [4] and Invasive Weed Optimization method (IWO) [5] have been proposed to address this problem of finding all the roots of given nonlinear equation or system of equations. In our earlier study, an interval based approach named as Novel GA has been proposed [6] to capture all the potential interval estimates for roots. In the present study, an "Intelligent Interval Labeling Algorithm" is proposed to identify these interval estimates into intervals containing roots and intervals containing extrema of given function.

M.N. Murty et al. (Eds.): MIWAI 2014, LNAI 8875, pp. 1–12, 2014.
© Springer International Publishing Switzerland 2014

GA are simulation programs [7] that create an environment which would allow only the fittest population to survive. Thus, GA provides some possible solutions and these solutions are represented as chromosomes. Generally the parameter 'Length of Chromosome' (LC) in GA is arbitrarily chosen and this might sometimes effect the performance of GA. Novel GA methodology [6] proposed earlier has two algorithms, namely *Pre-processing algorithm* and *Post-processing algorithm*. The fitness function in GA is defined as the sum of all potential intervals in the region of interest satisfying the condition that the product of function values at the end points of each interval is less than zero. To establish the knowledge about the number of roots that a nonlinear equation possesses, *Pre-processing algorithm* is proposed to suggest a mechanism that adaptively fixes the parameter LC in GA. GA coupled with *Post-processing algorithm* is executed with this new LC as input. Thus, the methodology [6] produces significantly narrow and potential interval estimates facilitating root computation to be highly efficient. Additionally it is noticed that Novel GA [6] has the power of capturing all the interval estimates i.e., missing a root in the zone of interest has less probability.

To address the problem of finding roots with multiplicity, several researchers attempted to address multiple zeros of a nonlinear equation [8-12]. The approach suggested by [8] has been adapted in Novel GA method [6]. It has the following transformation that converts roots with multiplicity of the problem $f(x) = 0$ as a problem $g(x) = 0$ with simple roots where

$$g(x) = \begin{cases} \dfrac{f(x)}{f'(x)}, & f'(x) \neq 0 \\ 0, & f'(x) = 0 \end{cases} \tag{1}$$

Hence in Novel GA method [6], $g(x)$ is considered for the initial population in the place of $f(x)$ in *Pre-processing algorithm* and we obtain the potential interval estimates of g. It is observed that all these selected potential intervals need not possess a root of f. By the nature of the smoothness of the function f, in between two roots i.e., in between two crossings of X-axis, there should be a maxima or minima of f. This paper aims at classifying each of these interval estimates into interval containing a root or interval having extrema of f. King [12] suggested that multiplicity $'m'$ of a root is approximately the reciprocal of the divided difference for $'g'$ for successive iterates. In the present study, a schematic method named as SGA is proposed to produce two conjugate pair of sequences, one is increasing and the other is decreasing and these sequences converge to a point in the interval. PIQA is developed by considering the function values of these two sequences, their absolute maximum, absolute difference and absolute relative difference between each pair of numbers in the two sequences.

The paper is organized as follows. Section 2 describes about the trisection method based Sequence Generating Algorithm (SGA). Section 3 explains about the Potential Interval Qualifier Algorithm (PIQA). Section 4 provides few illustrations and Section 5 has numerical examples and Section 6 concludes the results.

2   Sequence Generating Algorithm (SGA)

Consider the function $f(x) = (x - 1)^2 (\sin x)^2 + (x - 1)^3 (\cos x)^3 + 5(x - 1)$ that has 3 roots in $[0,4]$. Novel GA [6] is executed for the transformed function $g(x)$.

Table 1. The following 5 potential intervals are given as output from Novel GA [6]

S.No.	x-lower	x-upper
1	0.9776173	1.0271712
2	2.3490321	2.4026781
3	3.2538517	3.2985591
4	3.3498671	3.3588103
5	3.4281167	3.4696456

It is observed that 1st, 3rd and 5th intervals possess roots 1, 3.27577, 3.433599, whereas 2nd and 4th intervals do not possess any root of the given function f but these two intervals contain extrema of f. Thus, it is necessary to distinguish the intervals containing roots from the intervals containing extrema.

Let $[a, b]$ be a potential interval given as output from Novel GA [6] with respect to the function g such that $g(a) * g(b) < 0$. Generate two sequences $\{a_i\}, \{b_i\}, i = 0,1,..,n$ contained in $[a, b]$ with the only restriction that $\{a_i\}$ is an increasing sequence and $\{b_i\}$ is a decreasing sequence, $a_0 = a$ and $b_0 = b$. Thus, these two sequences have points that are uniformly densed. Now compute the absolute maximum, absolute difference and absolute relative difference defined by,

$$\text{Absolute maximum } h_i = Max\{|g(b_i)|, |g(a_i)|\},$$

$$\text{Absolute difference } d_i = |g(b_i) - g(a_i)|,$$

$$\text{Absolute relative difference } r_i = \frac{|g(b_i) - g(a_i)|}{b_i - a_i}, \text{ for } i = 0,1,2..,n$$

Interesting results are obtained as depicted in Fig. 1 and Fig. 2 for the sequences in the 1st and 2nd intervals given in Table 1.

Two sequences $\{a_i\}, \{b_i\}$ that are increasing and decreasing respectively are generated in the interval $[0.97762, 1,0272]$ with uniformly densed points . It can be noticed from Fig.1 (subplot 1) that function values of both the sequences are

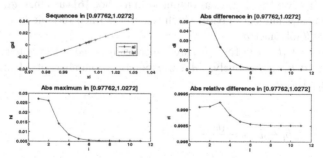

Fig. 1. Plot depicting the behavior of $\{a_i\}, \{b_i\}$ in 1st interval $[0.97762, 1,0272]$

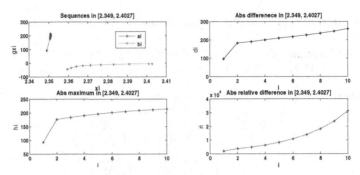

Fig. 2. Plot explaining the behavior of $\{a_i\}, \{b_i\}$ in 2nd interval $[2.349, 2.4027]$

converging in this interval. Absolute maximum, absolute difference monotonically decrease and are converging. Further, absolute relative difference also converges.

It can be seen from Fig.2 that in the interval $[2.349, 2.4027]$, function values of both the sequences are diverging, absolute maximum and absolute difference are monotonically increasing and absolute relative difference exponentially increases. Due to the transformation (1), $g(x)$ will be zero when it has root or when $f'(x) = 0$, which corresponds to the maxima or minima of f. Therefore, it can be understood that this interval encloses a point at which $f'(x) = 0$.

The above Fig. 1 and Fig. 2 illustrating the behavior of sequences (having uniformly densed points) do not explicitly explain the nature of the function f and function g in the interval. Hence, it is required to develop a schematic method to generate conjugate pair of sequences $\{a_i\}, \{b_i\}$ in $[a, b]$ such that always every pair $[a_i, b_i] \subseteq [a, b] \forall i = 0,1,2,\ldots,n$ maintains the potential interval property satisfying the condition $g(a_i) * g(b_i) < 0$ and every $[a_i, b_i]$ brackets a root. Motivated by the trisection method, the following SGA is developed for generating such conjugate pair of sequences.

Algorithm 1. *Sequence Generating Algorithm - SGA*

//Generates two sequences $\{a_i\}$, $\{b_i\}, i = 0,1,2..n$ *in* $[a, b]$ *satisfying potential interval property//*

INPUT:
- Given transformed function $g(.)$
- Potential interval $[a, b]$ given as output from method [6] such that $ga * gb < 0$
- p // p indicates number of parts which is a user defined variable //
- $\in \leftarrow 10^{-6}$ // tolerance //

OUTPUT: $[\{a_i\}, \{b_i\}, n]$ //$\{a_i\}, \{b_i\}$ are increasing and decreasing sequences contained in $[a, b]$ respectively. $'n'$ corresponds to number of elements in the sequences $\{a_i\}, \{b_i\}$ //

METHOD:

Step 1:
$$i \leftarrow 0, a_i \leftarrow a, b_i \leftarrow b, \delta \leftarrow \frac{(b-a)}{p}$$

Step 2:
   while (ä $>\in$)
      $i \leftarrow i + 1, anew \leftarrow a + \delta, bnew \leftarrow b - \delta$

compute $ganew \leftarrow g(anew), gbnew \leftarrow g(bnew)$
$Case$
$: ganew * gbnew < 0$
$a \leftarrow anew, b \leftarrow b_{new}, ga \leftarrow ganew, gb \leftarrow gbnew$
$Case$
$: ga * ganew < 0$
$b \leftarrow anew, gb \leftarrow ganew$
$Case$
$: else$
$a \leftarrow bnew, ga \leftarrow gbnew$
$end\ Case$
$a_i \leftarrow a, b_i \leftarrow b$
$\delta \leftarrow \frac{(b-a)}{p}$
$end\ while$
$n \leftarrow i$
End; //SGA //

3 Potential Interval Qualifier Algorithm (PIQA)

The following result is given by [8, p5].

Lemma: If the given function $f(.)$ has a root $'\alpha'$ with multiplicity $'m'$, and $g(.)$ is the transformed function by (2), then

$$\lim_{x \to \alpha} \frac{g(x)}{(x - \alpha)} = \frac{1}{m}, \text{i.e., if } \delta = x - \alpha \Rightarrow x = \alpha + \delta \text{ and as } x \to \alpha, \delta \to 0, \text{then}$$

$$\lim_{x \to \alpha} \left\{ \frac{1}{\delta} \frac{f(\alpha + \delta)}{f'(\alpha + \delta)} \right\} = \frac{1}{m}$$

King [9] proved that multiplicity $'m'$ of a root is approximately equal to the reciprocal of the divided difference for $'g'$ for successive iterates. We now state the following theorem.

Theorem: Given two sequences $\{a_i\}, \{b_i\}$ in $[a, b]$ generated by SGA such that both the sequences converge to root α with multiplicity $'m'$. Let $a_i = \alpha - \delta_i, b_i = \alpha + \delta_i$ for $i = 0, 1, 2.., n$. Then the absolute relative difference

$$\lim_{\delta_i \to 0} \left\{ \frac{|g(b_i) - g(a_i)|}{b_i - a_i} \right\} = \frac{1}{m}.$$

Now, the following Potential Interval Qualifier Algorithm is proposed.

Notation:
- $[a^j, b^j]$ denotes j^{th} interval
- $\{a_i^j\}, \{b_i^j\}$ are monotonically increasing and decreasing sequences generated by SGA that confine to j^{th} interval.

Algorithm 2. PIQA

//This algorithm qualifies the given potential interval estimates in to intervals enclosing root (with multiplicity) and intervals containing extrema //

INPUT:

- Given transformed function $g(.)$
- $d \leftarrow$ Number of potential intervals $[a^j, b^j], j = 1,2..d$ given as output from Novel GA method [6] such that $g(a^j) * g(b^j) < 0$, for each j
- p // p indicates number of parts which is a user defined variable //
- $\in \leftarrow 10^{-6}$ // tolerance to generate sequences in SGA//
- $\in_1 \leftarrow 10^{-3}$// tolerance to qualify the interval whether it has a root or extrema //
- $[\{a_i^j\}, \{b_i^j\}, n^j], j = 12, ...d$ //output from SGA where n^j corresponds to number of terms in the pair of sequences $\{a_i^j\}, \{b_i^j\}$ generated in j^{th} interval//

OUTPUT: $[\{h_i^j\}, \{d_i^j\}, \{r_i^j\}, pi, npi, m]$ //$\{h_i^j\}, \{d_i^j\}, \{r_i^j\}$ correspond to sequences for absolute maximum, absolute difference, absolute relative difference respectively in j^{th} interval. $'pi'$ corresponds to number of potential intervals possessing root in the region of interest, $'npi'$ corresponds to number of intervals that have extrema but do not contain root and $'m'$ is an array representing multiplicity of each root in the corresponding interval //

METHOD:

Step 0: Initialization: $j \leftarrow 1, pi \leftarrow 0, npi \leftarrow 0$

Repeat Step 1 through Step 4 *while* $(j \leq d)$

Step 1: Initialize j^{th} interval $[a^j, b^j]$

Step 2: $[\{a_i^j\}, \{b_i^j\}, n^j] \leftarrow SGA(g(.), [a^j, b^j], p, \in)$

Step 3: Compute *for* $i \leftarrow 1$ *to* n^j

$$h_i^j \leftarrow Max\{|g(a_i^j)|, |g(b_i^j)|\}$$

$$d_i^j \leftarrow |g(b_i^j) - g(a_i^j)|$$

$$r_i^j \leftarrow \frac{|g(b_i^j) - g(a_i^j)|}{b_i^j - a_i^j}$$

// sequences for absolute maximum, absolute difference, absolute relative difference in the interval $[a^j, b^j]$ //

Compute $Min\_hij \leftarrow min (h_i^j); Min\_dij \leftarrow min (d_i^j); Min\_rij \leftarrow min (r_i^j)$
          endfor

Step 4: $if((\ (Min\_hij <\in_1\ )\ and\ (Min\_dij <\in_1))$
               $pi \leftarrow pi + 1$
         $m(j) \leftarrow callmr(\{r_i^j\}, \in_1, n^j)$ //Returns multiplicity of each root//
               else
             $npi \leftarrow npi + 1$
       endif

   endwhile

End; *//PIQA //*

Algorithm to find multiplicity: $callmr(\{r_i^j\}, \in_1, n^j)$

$$l \leftarrow n^j$$
$$if((r_i^j - r_{i-1}^j) < \in_1)$$
$$m = \frac{1}{r_i^j}$$
$$endif$$

4   Illustration for SGA and PIQA

Now consider $f(x) = (x-1)^2 (\sin x)^2 + (x-1)^3 (\cos x)^3 + 5(x-1)$. While generating sequences by SGA, we come across three cases.

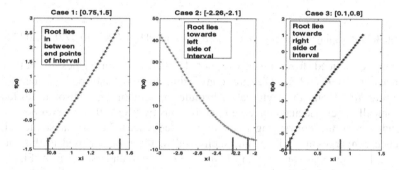

Fig. 3. Plot depicting the three possible cases in three different intervals

In the above Fig. 3, subplot 1 shows the behavior of function in the interval [0.75,1.5], where root lies inside the interval satisfying the condition that product of function values at the end points of the interval is less than zero. Subplot 2 exhibits the case 2 where root lies outside the interval towards left side. Subplot 3 displays the case 3 where root lies outside the interval towards right side. Now, while generating two sequences inside a given interval by SGA, we always wish to bracket the root. Thus, the two sequences $\{a_i\}, \{b_i\}$ are generated in such a way that always every pair $[a_i, b_i] \subseteq [a, b] \, \forall i = 0,1,2,\ldots,n$ maintains the potential interval property satisfying the condition $g(a_i) * g(b_i) < 0$. The consecutive new values in each pair of sequences are altered accordingly to satisfy this condition as given in Algorithm 1. Hence this is a learning algorithm, a knowledge discovery approach by means of extracting the information in the given interval. Further, if the interval encloses a root, then function values of f and g for both the sequences converge to zero in the interval and hence both the sequences will converge to the root. If the potential interval does not contain a root, then the function values of f for both the sequences diverge whereas function values of g for both the sequences will have erratic behavior. As the transformed function g possesses quality information, the methodology is developed and analysis is made with focus on the sequences of function values of g, their absolute maximum, absolute difference and absolute relative difference.

It is understood that if δ is too small, we get refined sequences that are always falling into case 1 while generating them. In general, taking a smaller delta influences resource utilization. It is noticed that the potential interval estimates returned by Novel GA are very narrow and hence it is not required to have such smaller δ in SGA.

Now generate sequences $\{a_i\}$, $\{b_i\}$ by SGA in each of the potential intervals given in Table 1 and apply PIQA that qualifies each potential interval into interval enclosing root or interval containing extrema.

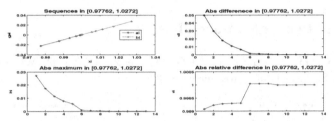

Fig. 4. Plot displaying the behavior $\{a_i\}$, $\{b_i\}$ generated by SGA in $[0.97762, 1,0272]$

It can be noticed from Fig.4 (subplot 1) that function values of both these sequences in $[0.97762, 1,0272]$ are converging to zero and both sequences are converging to '1' in this interval. Absolute maximum and absolute difference monotonically decrease and converge to zero. Thus by PIQA, this interval possesses a root and root is the converging point of two sequences. Further, absolute relative difference converges to 1, i.e., multiplicity $'m'$ of the root enclosed in this interval equals to 1. In comparison with Fig.1, it can be clearly understood from Fig. 4 that this algorithm qualifies whether a given interval possesses a root and if it has a root, its multiplicity is also obtained.

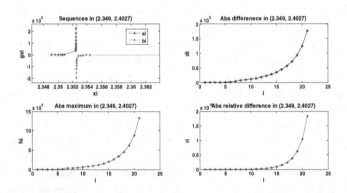

Fig. 5. Plot depicting the behavior of $\{a_i\}$, $\{b_i\}$ generated by SGA in $[2.349, 2.4027]$

In contrast to the Fig.2, it can be clearly seen from Fig.5 that in the interval $[2.349, 2.4027]$, function values of both the sequences are diverging, absolute maximum and absolute difference are monotonically increasing and absolute relative difference exponentially increases. Hence it can be concluded from PIQA that this interval does not possess root but it has extrema of f.

5   Numerical Experiments

In this section we present 5 examples to illustrate the efficiency of the proposed method.

Table 2. A set of bench mark functions and the results with the proposed method are listed below. Here in this table, NI represents number of intervals returned by Novel GA [6], NIR corresponds to number of intervals enclosing root and NIE corresponds to number of intervals having extrema. It can be noted that 'm' in 8th column denotes the multiplicity of root.

Function	Zone	NI by [6]	NIR	NIE	No. of selected roots		
					total	simple	multiple
$f_1 = (x(e^{x^2}) - (sinx)^2 + 3cosx + 5)^4$	[-3,4]	3	1	2	4	-	1 $(m = 4)$
$f_2 = sin(0.2x)cos(0.5x)$	[-50,50]	37	19	18	23	15	4 $(m = 2)$
$f_3 = (3x - 2)^4(2x - 3)^2 (96x^3 - 332x^2 + 325x - 75)$	[0.5,2]	7	4	3	8	2	2 $(m_1 = 4)$
$f_4 = \frac{2}{3} - -(0.1 - x^{11})e^{2-x^2}$	[-1,1]	5	3	2	3	3	-
$f_4 = (64x^4 - 16\pi x^3 - 3\partial^2 x^2 + \pi^3 x - \frac{\partial^4}{16})(sin5x + 0.5x + 2)$	[-1,1]	7	3	4	4	2	1 $(m = 2)$

Consider f_2 that has total 23 roots in $[-50,50]$ among which 19 are distinct roots.

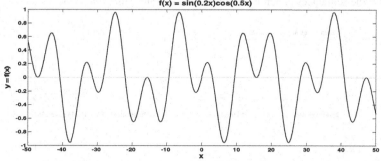

Fig. 6. Plot for $f_2 = sin(0.2x)cos(0.5x)$

As shown in Fig 6, f_2 has 15 simple roots and 4 roots are multiple roots, each with multiplicity 2. Novel GA [6] returned 37 narrow potential intervals and SGA along with PIQA qualifies these 37 intervals into 19 intervals enclosing a root of both g as well as f_2 and 18 intervals as intervals containing extrema of f_2. The following Table 3 displays the intervals enclosing roots and intervals containing extrema of the function f_2.

Table 3. Intervals possessing root or extrema of f_2

S. No.	x-lower	x-upper	PIQA output	S. No.	x-lower	x-upper	PIQA output
1	-47.3684606	-46.9983821	Root	20	1.6432410	1.7254402	Extrema
2	-43.2214511	-43.1660531	Extrema	21	3.125396	3.1685314	Root
3	-40.8441857	-40.8193238	Root	22	6.426091	6.5067159	Extrema
4	-37.9821880	-37.8580232	Extrema	23	9.323579	9.7045229	Root
5	-34.5748749	-34.4447411	Root	24	11.539284	11.915735	Extrema
6	-33.2650959	-33.0479707	Extrema	25	15.684648	15.772927	Root
7	-31.5959871	-31.3240455	Root	26	19.530681	19.656868	Extrema
8	-29.7383903	-29.6495947	Extrema	27	21.746960	22.132692	Root
9	-28.3528603	-28.2703011	Root	28	24.652562	25.024529	Extrema
10	-24.9694113	-24.8496210	Extrema	29	28.108451	28.296819	Root
11	-22.1215471	-21.9769910	Root	30	29.681578	29.814171	Extrema
12	-19.8794413	-19.5596421	Extrema	31	31.377958	31.623528	Root
13	-15.7578597	-15.6994188	Root	32	32.917715	33.118862	Extrema
14	-11.9152470	-11.7062010	Extrema	33	34.545802	34.775856	Root
15	-9.5616142	-9.3720057	Root	34	37.832769	37.948854	Extrema
16	-6.6704734	-6.3067219	Extrema	35	40.605173	40.843837	Root
17	-3.1601239	-2.8432123	Root	36	43.161613	43.511411	Extrema
18	-1.7326985	-1.2000641	Extrema	37	47.014463	47.584481	Root
19	-0.2161691	0.09399237	Root	-	-	-	-

It can be observed that from Table 3 and Fig.6 that the output given by PIQA matches with the roots and extrema of f_2. The following Fig 7 and Fig 8 depict the convergence criterion for the sequences generated in the 1st and 4th intervals from Table 3.

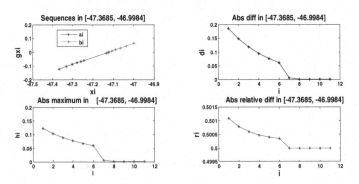

Fig. 7. Plot describing the behavior of $\{a_i\}, \{b_i\}$ generated by SGA in $[-47.3685, -46.9984]$

The function f_2 has a root in the interval $[-47.3685, -46.9984]$ with multiplicity 2. The behavior of the two sequences generated by SGA in this interval can be seen in Fig.7 that function values of both the sequences, absolute maximum and absolute difference converge to zero. Hence by PIQA, it can be concluded that this potential interval contains a root. It can also be observed from Fig.7 (subplot 4) that absolute relative difference converges to 0.5 and hence the multiplicity of the root enclosed in this interval is reciprocal of 0.5, which equals to 2.

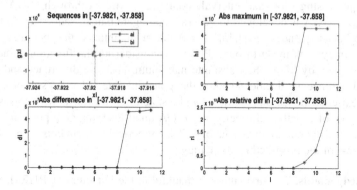

Fig. 8. Plot displays the behavior of $\{a_i\}, \{b_i\}$ generated by SGA in $[-37.9821, -37.858]$

The sequences for function values are diverging, absolute maximum and absolute difference are monotonically increasing. Further, sequence for absolute relative difference is exponentially increasing. Hence, it can be concluded that by PIQA that this interval contains extrema.

Now consider the function $f_3 = (3x - 2)^4 (2x - 3)^2 (96x^3 - 332x^2 + 325x - 75)$ that has 4 distinct roots in $[0.5, 2]$, among which 0.6666 is a root with multiplicity 4.

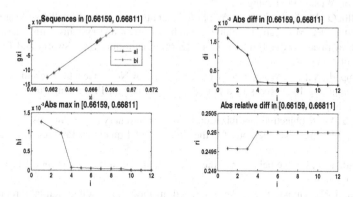

Fig. 9. Plot depicting the behavior of $\{a_i\}, \{b_i\}$ generated by SGA in $[0.66159, 0.66811]$

As the sequences for absolute maximum, absolute difference converge to zero, it can be stated that this potential interval encloses a root and subplot 4 explains that this root (0.6666) has multiplicity 4.

6   Conclusion

A schematic method is proposed to classify the given potential interval estimates into intervals possessing roots and intervals containing extrema through "SGA" coupled with "PIQA". SGA is developed to generate conjugate pair of sequences so that each pair $[a_i, b_i]$ of sequences $\{a_i\}, \{b_i\}, i = 0,1,2..n$ maintains the potential interval property. Analysis is made through PIQA by considering the function values of the sequences given by SGA, their absolute maximum, absolute difference and absolute relative difference to decide whether the potential interval possesses a root or it contains extrema. Based on [8,12], it can be stated that root of g with corresponding $m \geq 1$ is root of f with multiplicity $'m'$. PIQA qualifies remaining potential intervals of g as intervals containing extrema of f. The numerical experiments demonstrate the effectiveness of the proposed methodology.

Acknowledgements. The first author is thankful to the Director, ACRHEM, UoH and to the Director, HEMRL, Pune for the financial support.

References

1. Numerical Recipes: The Art of Scientific Computing, 3rd edn. Cambridge Univ. Press
2. Holland, J.H.: Adaption in natural and artificial systems: An introductory analysis with applications to biology, control, and artificial intelligence. University of Michigan Press
3. Goldberg, D.E.: Genetic Algorithms in Search, Optimization and Machine Learning. Addison-Wesley Publishing
4. Castillo, O., Melin, P., Pedrycz, W.: Soft Computing for Hybrid Intelligent Systems
5. Pourjafari, E., Mojallali, H.: Solving nonlinear equations systems with a new approach based on Invasive weed optimization algorithm and clustering. Swarm and Evolutionary Computation 4, 33–43 (2012)
6. Nadimpalli, V.L.V., Wankar, R., Chillarige, R.R.: A Novel Genetic Algorithmic Approach for Computing Real Roots of a Nonlinear Equation. In: Proceedings of EVO Star 2014. LNCS (in press, 2014)
7. Deb, K.: Multi-objective Optimization Using Evolutionary Algorithms. Wiley (2001)
8. Traub, J.F.: Iterative Methods for the Solution of Equations. Prentice Hall, Englewood (1964)
9. Espelid, T.O.: On the behavior of the secant method near a multiple root. BIT I1, 112–115 (1972)
10. Parida, P.K., Gupta, D.K.: An improved method for finding multiple roots and it's multiplicity of nonlinear equations in R*. Applied Mathematics and Computation 202, 498–503 (2008)
11. Yun, B.I.: Iterative methods for solving nonlinear equations with finitely many zeros in an interval. Journal of Computational and Applied Mathematics 236, 3308–3318 (2012)
12. King, R.F.: A secant method for multiple roots. BIT 17, 321–328 (1977)

Stochastic Leaky Integrator Model for Interval Timing

Komala Anamalamudi[1,2], Bapi Raju Surampudi[1,3],
and Madhavilatha Maganti[4]

[1] School of Computer and Information Sciences, University of Hyderabad, India
[2] Madanapalle Institute of Technology & Science, Madanapalle, India
[3] Cognitive Science Lab, IIIT-Hyderabad, India
[4] Center for Neural and Cognitive Sciences, University of Hyderabad, India
komal_nag@yahoo.com, bapics@uohyd.ernet.in, magantimadhavilatha@gmail.com

Abstract. Interval timing plays an important role in every aspect of our daily life. Intelligent behavior, adaptation and survival of mankind rely on proper judgment of temporal intervals. Since many decades, Pacemaker Accumulator Model (PAM) has been the most influential interval timing model in human time perception domain. It is purely a psychological model and many researchers from the neurobiology domain attempted to find the biological equivalents of the components of PAM. In this paper, we propose a computational model of interval timing based on spiking neurons which is biologically plausible yet preserving the simplicity and strength of psychological models. Preliminary results demonstrate that the computational model we proposed can mimic empirical data from psychological experiments.

Keywords: Cognitive modeling, Spiking neural networks, Time perception, Leaky integrate-and-fire, Scalar property.

1 Introduction

"Timing is everything: in making shots, in making love, in making dinner. Indeed, it is difficult to conceive of an action that doesn't require temporal control...In addition to coordinating complex sequences, timing serves a very basic function long recognized as a fundamental aspect of the learning process in animals: anticipation or prediction..." [8]. Time Perception refers to sense of time that involves experience and passage of time, temporal discrimination and time estimation. Time Perception is widely studied in Cognitive Science domains from varied perspectives. For example, psychologists study time perception as part of behavioral analysis [4] [28]; neuroscientists study time perception as part of finding information processing mechanisms in brain [12][10] and computer scientists study time perception as part of computational modeling and development of autonomous intelligent systems [15].

Multiple neural systems are responsible for processing multiple time scales in at least 10 orders of magnitude [17]. Buhusi and Meck [2] classify different orders

M.N. Murty et al. (Eds.): MIWAI 2014, LNAI 8875, pp. 13–22, 2014.

of time into *circadian timing, interval timing* and *millisecond timing*. The circadian clock that keeps track of sleep-wake cycle, appetite and behavioral rhythms exists in suprachiasmatic nucleus (SCN) [12]. The millisecond timer responsible for speech processing and motor coordination exists in cerebellum [2]. The time scale in the range of seconds to minutes range is referred to as interval timing and involves conscious time estimation. The neural mechanisms of interval timing are not clearly identified and Ivry and Schlerf specify that some models are based on dedicated mechanisms and some models are based on intrinsic mechanisms [10]. Some psychologists categorize timing mechanisms into automatic and cognitively controlled mechanisms. The circadian timing and millisecond timing are labeled as automatic timing mechanisms and interval timing is labeled as cognitively controlled timing.

There are four standard psychophysical investigation methods for studying time perception: *verbal estimation, interval reproduction, interval production* and *interval comparison. Verbal estimation* refers to specifying the duration of a stimulus in time units (usually in seconds) ; *interval reproduction* refers to reproducing the duration of the presented stimulus by the repetition of same event or by other means. Some people refer to time reproduction task as *peak-interval procedure* [16], as responses from several trials peak at the criterion duration. *Interval production* refers to producing a timing task of the given duration; *interval comparison* refers to comparing different lengths of the same stimulus or comparing lengths of two different stimuli [28].

Interval timing exhibits an important characteristic namely *scalar property*. For both psychological and neural models of interval timing, accountability is achieved by *scalar property. Scalar property*, a specialized form of Weber's law, specifies that the time varying behavior of the subject stretches or scales in proportion to the length of the duration of the stimulus. In other words, it refers to the linear relationship between the standard deviation of the time estimate and mean of time estimate. Treisman [24] defines *scalar property* as the phenomenon wherein the error in time estimation increases with increased duration of the stimulus.

The cognitive and neural models of time perception are broadly categorized into dedicated and intrinsic models [10][27] as shown in Table 1 . Computational models are feasible for comprehensive understanding of mind and brain. In time perception realm, there is a lacunae for computational modeling. There are a very few computational models of interval timing [23] [1] [19]. The computational model by Taatgen et al. [23] is a procedural model and is built on ACT-R architecture. Addyman et al. [1] devised and simulated a computational model for interval timing in infants using a connectionist memory-decay approach. Oprisan and Buhusi [19] devised a mathematical model to analyze the characteristics of cosine oscillators and the role of noise in striatal beat frequency model.

We propose Stochastic Leaky Integrator Model(SLIM) which is based on leaky integrate-and-fire spiking neuron model and the idea is inspired by PAM. Our model, SLIM is a hybrid model that combines the features of dedicated and intrinsic models of timing. It combines a biologically plausible model of neuron

Table 1. Models of Time Perception

Dedicated Models	Intrinsic Models
- In dedicated models, a neural structure or components for handling temporal information could be a specialized one or a distributed network of different brain regions or components. - Cerebellum, basal ganglia, supplementary motor area and right prefrontal cortex are the examples of specialized dedicated mechanisms proposed in various neural models of time perception. - PAM is the seminal cognitive model that assumes a dedicated clock component to deal with time - Other examples of dedicated models are Beat Frequency model by Miall [18] and Connectionist Model proposed by Church and Broadbent [5]	- In intrinsic models, timing is inherent and is part of sensory information processing. - State dependent network models and energy readout models come under this category. - Other examples are Memory-decay model proposed by Staddon [22], Dual Klepsydra model by Wackermann and Ehm [26], Population clocks proposed by Buonomano and Laje [3], Spectral timing model by Grossberg and Schmajuk [9].

activation with counting mechanism in order to handle interval timing as cortical neurons fire in the range of milliseconds. SLIM is a simple, biologically plausible computational model for interval timing. Section 2 presents an overview of spiking neural networks and delineates the feasibility of leaky integrate-and-fire neuron model as the computational substrate of interval timing.

2   Spiking Neural Networks

A neuron is assumed to be a dynamic element that generates pulses or spikes whenever its excitation reaches a threshold value. Biological neural systems use spike timing for information processing. The generated sequence of spikes contains the information and it gets transmitted to the next neuron [7]. Spiking neurons are the computational units in brain. The presence of output spikes from the neurons and/or the timing of the spiking neurons is assumed to be the information transmission mechanism in spiking neural networks. Though artificial neural networks are proved to be powerful problem solving tools in the domains such as bioinformatics, pattern recognition, robotics etc., they suffer in processing large amounts of data and adaptation to the dynamic environment [20]. Maass [14] quotes spiking neural networks as third generation of artificial neural networks, the first generation of networks being threshold logic units and the second generation of neural networks being sigmoidal units (see Table 2).

Table 2. Classification of Neural Networks according to [14]

Generation	Computational Unit	Working Principle	Architecture/ Model
First Generation	Threshold Gates (McCulloch-Pitts neurons)	Digital output	Perceptrons, Hopfield Network, Boltzmann machine
Second Generation	Sigmoidal Units	Real valued outputs which are interpreted as firing rates of natural neurons	Multilayer perceptrons, Radial Basis Function networks
Third Generation	Spiking Neurons	Spike Timing	Hodgkin-Huxley model, Integrate-and-Fire model, Spike response model

2.1   Spiking Neuron Models

The most influential spiking neuron models are: Hodgkin-Huxley Model, Integrate-and-Fire Model and Spike Response Model. Hodgkin-Huxley Model is a traditional conductance-based model of spiking neurons. When compared to Integrate-and-Fire model and Spike response model, it is quite complex to analyze mathematically. Spike Response model could be demonstrated as a generalized version of Integrate-and-Fire model. Leaky integrate-and-fire model is a variant of integrate-and-fire model wherein , a neuron is modeled as a leaky integrator of its input. The neurons in integrate-and-fire model could be stimulated by some external current or by the input from presynaptic neurons.

We focus on leaky integrate-and-fire model as the nature of leaky integration of spikes resembles the accumulation of pulses in accumulator of PAM. And, when the membrane potential reaches a threshold, an output spike is generated and the no.of spikes that gets generated is in proportion to the length of the stimulus.Eq. 1 describes a simple resistor-capacitor (RC) circuit where the neuron is modeled as a leaky integrator with the input I(t).

$$\tau_m \frac{d\nu}{dt} = -\nu(t) + RI(t) \qquad (1)$$

The membrane potentials of the leaky integrate-and-fire neurons are calculated depending on the input type. Eqs.2, 3, 4 describe the computation of membrane potential with the inputs of constant current, time dependent stimulus and presynaptic currents respectively.

1. Stimulation by a constant input current I(t):

$$\nu(t) = RI[1 - \exp\left(-\frac{t}{\tau_m}\right)] \qquad (2)$$

2. Stimulation by a time-varying input current:

$$\nu(t) = \nu_r \exp\left(-\frac{t - t_0}{\tau_m}\right) + \frac{R}{\tau_m} \int_0^{t-t_0} \exp\left(-\frac{s}{\tau_m}\right) I(t - s)ds \qquad (3)$$

3. Stimulation by synaptic currents:

$$I_i(t) = \sum_j w_{ij} \sum_f \alpha(t - t_j^{(f)}) \qquad (4)$$

For devising the model of interval timing, we considered the case of 'stimulation by synaptic currents', as the sensory stimulus passes through the cortical neurons and cortical neurons would act as presynaptic neurons for leaky integrate-and-fire neuron.

3 The Model

After an extensive review of literature on time perception, we identified that there is a lacunae of computational models of time perception. To fill this gap in literature, we attempted to devise a computational model that is biologically plausible based on spiking neurons. In addition, the model should satisfy the criterion of being lucid and concise for explaining the mechanisms involved in interval timing like PAM. This led to devising a hybrid model possessing the properties of dedicated and intrinsic models of time perception. The model is called Stochastic Leaky Integrator Model (SLIM) because it depicts the stochastic firing nature of cortical neurons and leaky integration of the integrate-and-fire neuron. The schematic representation of SLIM is presented in Fig. 2.

In PAM (Fig. 1), which is based on Scalar Expectancy Theory (SET), a Poisson *pacemaker* continuously emits pulses. The *switch* is closed on the onset of a stimulus and the pulses get accumulated in *accumulator*. During training, the contents of the *accumulator* are stored in *reference memory*. During testing, the contents of the *accumulator* are stored in *working memory* and a response is generated as an outcome of the *ratio comparison* between contents of *working memory* and *reference memory*. When there is a sensory stimulus, the cortical neurons start firing stochastically at irregular intervals [6]. So, it is convenient to maintain the number of spikes/pulses generated by the cortical neurons during the presence of stimulus rather than keeping track of spiking times of the neurons over a time course. Our model SLIM exactly utilizes this property for interval time estimation. According to SLIM, when there is a stimulus, cortical neurons start firing and generate spikes. To mimic the random firing nature of cortical neurons, we used a uniformly distributed random number generator to determine the number of neurons that fire at any instance of time. During the presence of stimulus, the potential of the leaky integrate-and-fire (LIF) neuron increases in proportion to the spiking of presynaptic neurons. When the potential reaches a threshold, the LIF neuron generates an output pulse and counter keeps track of these pulses. After reaching threshold, potential is set to a resting value and the integration of potentials starts again at LIF neuron. This accumulation continues until the stimulus is presented and after that the contents of counter are shifted to memory and a response is generated(see Algorithm1).

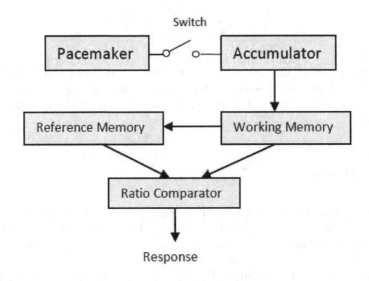

Fig. 1. Pacemaker Accumulator Model

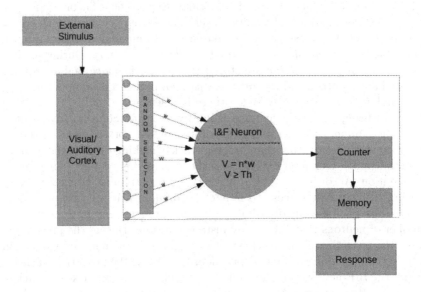

Fig. 2. Stochastic Leaky Integrator Model for Interval Timing

---

Algorithm 1. Functioning of SLIM

Network Architecture & Initializations:
No. of cortical neurons, M= 100
No. of Leaky integrate-and-fire neurons: 1
Threshold, Th= 50
Resting potential, R= 0.5
Presynaptic weight of each neuron, W= 0.75
Potential, V= 0
Counter, S= 0

1: **while** STIMULUS **do**
2:     When there is an external stimulus, the cortical neurons start firing
3:     The no. of cortical neurons that fire at a moment are selected by a random number generator;
4:     N= Random(M)
5:     The potential of leaky integrate-and-fire neuron is calculated as the weighted sum of firing neurons.
6: $V = N_*W$
7:     **if** $V \geq Th$ **then**
8:         LIF neuron generates a spike and counter is incremented by one
9:         $S = S + 1$
10:        Potential, V is set to resting potential R
11:        $V = R$
12:     **else**
13:        $I = N_*W$
14:        $V = V + I$
15:     **end if**
16: **end while**
17: Shift the values of Counter to Memory;
18: $Mem = S$
19: **return** S

---

4   Results and Discussion

We experimented to model the results of time reproduction task given in [21] for 8s, 12s and 21s through SLIM. Time duration is plotted on x-axis and the percentage of response rate is plotted on y-axis. We simulated the model with 100 presynaptic neurons and the number of neurons that get excited at an instance of time is chosen by a uniformly distributed random number generation function to demonstrate more realistic the firing of neurons. Fig. 3 shows the result of the simulations with *fixed threshold*. To mimic the results of behavioral tests done by Rakitin et al. [21], we also ran the simulations for 80 trials and the outcomes of these trials are plotted in Fig. 3.

Though there is a significant variance between results of Rakitin et al. experiments and the results of simulations of SLIM, we welcome the stochastic firing nature of cortical neurons as it is the mundane nature of biological neurons. Another source of variance is *fixed threshold*. We considered a threshold value of

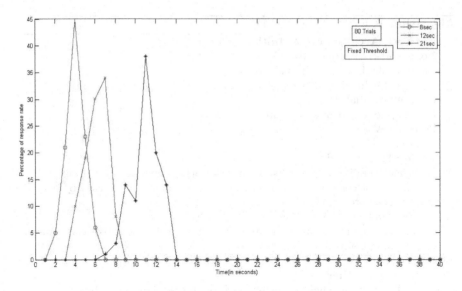

Fig. 3. Performance of SLIM with fixed threshold

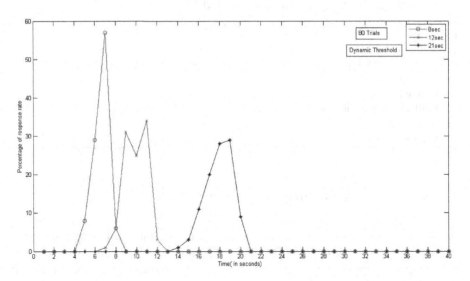

Fig. 4. Performance of SLIM with dynamic threshold

50 as it is a quite reasonable assumption that atleast 50% neurons should fire in order to generate an output spike.

To reduce variance between SLIM results and Rakitin et al. results, we attempted to work with *dynamic threshold*. Initially the threshold of the simulations start with 50 and later the threshold is reduced by the inhibitory input at previous instance. The results of the simulations are much improved by inducing *dynamic threshold* which is again a common feature of biological neurons [11][13] and are shown in Fig. 4. As we focused much on studying the behavior of stochastic leaky integrator as a computational substrate of interval timing that is also biologically plausible, at present we did not work with training the model. We hope that if the model is trained with a reasonable number of trials, the results of the simulations of SLIM would closely resemble the results given by Rakitin et al.

5   Conclusion and Future Work

The model proposed in this paper, SLIM is simple yet computationally efficient. Further, it integrates the features of dedicated and intrinsic models of time perception and these features are represented by spiking cortical neurons and the counting mechanism respectively.

As the model is based on spiking neurons it is evident that the model is biologically plausible [25]. To make the model more robust and compatible with psychological models of time perception, the results from behavioral experiments by Rakitin et al. were used to assess the performance of SLIM. Initially the model was tested for 8s, 12s and 21s time durations with fixed threshold of leaky integrate-and-fire neuron. To improve the results, dynamic threshold that varies from iteration to iteration and whose value depends on inhibitory input of previous iteration is considered. Though the results of simulations using dynamic threshold are much better than the results of simulations using fixed threshold, the results indicated that there is no close resemblance to the results of Rakitin et al. experiments. This may be due to lack of training for the model.

Future line of work should focus on incorporating training for the model. It is assumed that by incorporating training, the model would be congruent with psychological models. Further it is intended to study the feasibility of spiking neuron populations for interval timing. At present, we did not focus on time course of spiking neurons, and in future we intend to explore it to study the relation between time course and interval timing.

References

1. Addyman, C., French, R.M., Mareschal, D., Thomas, E.: Learning to perceive time: A connectionist, memory-decay model of the development of interval timing in infants. In: Proc. in Livre/Confrence Cognitive Science (2011)
2. Buhusi, C.V., Meck, W.H.: What makes us tick? Functional and neural mechanisms of interval timing. Nat. Neurosci. 6, 755–765 (2005)
3. Buonomano, D.V., Laje, R.: Population clocks: motor timing with neural dynamics. Trends in Cognitive Sciences 14(12), 520–527 (2010)
4. Church, R.M.: A tribute to john gibbon. Behavioural Processes 57, 261–274 (2002)
5. Church, R.M., Broadbent, H.A.: Alternative representations of time, number, and rate. Cognition 37(1), 55–81 (1990)

6. Gerstner, W.: Integrate-and-fire neurons and networks. In: The Handbook of Brain Theory and Neural Networks, vol. 2, pp. 577–581 (2002)
7. Gerstner, W., Kistler, W. (eds.): Spiking Neuron Models: Single Neurons, Populations, Plasticity, Cambridge (2002)
8. Gibbon, J., Malapani, C.: Neural basis of timing and time perception. In: Nadel, L. (ed.) Encyclopaedia of Cognitive Science, London, UK (2001)
9. Grossberg, S., Schmajuk, N.A.: Neural dynamics of adaptive timing and temporal discrimination during associative learning. Neural Networks 2(2), 79–102 (1989)
10. Ivry, R.B., Schlerf, J.E.: Dedicated and intrinsic models of time perception. Trends in Cognitive Science 12(7), 273–280 (2008)
11. Izhikevich, E.M.: Which model to use for cortical spiking neurons? IEEE Transactions on Neural Networks 15(5), 1063–1070 (2004)
12. Karmarkar, U.R., Buonomano, D.V.: Timing in the absence of clocks: Encoding time in neural network states. Neuron 53(3), 427–438 (2007)
13. Kumar, S.: Neural Networks: A classroom approach. McGraw-Hill Higher Education (2011)
14. Maass, W.: Networks of spiking neurons: The third generation of neural network models. Neural Networks 10, 1659–1671 (1996)
15. Maniadakis, M., Trahanias, P.: Temporal cognition: A key ingredient of intelligent systems. Frontiers in Neurorobotics 5 (2011)
16. Matell, M.S., Meck, W.H.: Cortico-striatal circuits and interval timing: coincidence-detection of oscillatory processes. Cogn. Brain Res. 21(2), 139–170 (2004)
17. Mauk, M.D., Buonomano, D.V.: The neural basis of temporal processing. Annu. Rev. Neurosci. 27, 307–340 (2004)
18. Miall, C.: The storage of time intervals using oscillating neurons. Neural Computation 1(3), 359–371 (1989)
19. Oprisan, S.A., Buhusi, C.V.: Why noise is useful in functional and neural mechanisms of interval timing? BMC Neuroscience 14, 48 (2013)
20. Paugam-Moisy, H., Bohte, S.: Computing with spiking neuron networks. In: Handbook of Natural Computing, pp. 335–376. Springer (2012)
21. Rakitin, B.C., Gibbon, J., Penney, T.B., Malapani, C., Hinton, S.C., Meck, W.H.: Scalar expectancy theory and peak-interval timing in humans. Journal of Experimental Psychology: Animal Behavior Processes 24(1), 15–33 (1998)
22. Staddon, J.E.R.: Interval timing: memory, not a clock. Trends in Cognitive Sciences 9(7), 312–314 (2005)
23. Taatgen, N.A., Rijn, H.V., Anderson, J.R.: An integrated theory of prospective time interval estimation: The role of cognition, attention and learning. Psychological Review 114(3), 577–598 (2007)
24. Triesman, M.: Temporal discrimination and indifference interval: Implications for a model of the internal clock. Psychological Monographs 77(13), 1–31 (1968)
25. Vreeken, J.: Spiking neural networks, an introduction. Institute for Information and Computing Sciences, Utrecht University Technical Report UU-CS-2003-008 (2002)
26. Wackermann, J., Ehm, W.: The dual klepsydra model of internal time representation and time reproduction. Journal of Theoretical Biology 239(4), 482–493 (2006)
27. Wittmann, M.: The inner sense of time: how the brain creates a representation of duration. Nature Reviews Neuroscience 14(3), 217–223 (2013)
28. Zakay, D., Block, R.A.: Temporal cognition. Curr. Dir. Psychol. Sci. 6, 12–16 (1997)

Multi-objective Exploration for Compiler Optimizations and Parameters

N.A.B Sankar Chebolu[1,2] and Rajeev Wankar[2]

[1] ANURAG, Hyderabad, India
[2] Department of Computer and Information Sciences, University of Hyderabad,
Hyderabad, India
chnabsankar@yahoo.com, rajeev.wankar@gmail.com

Abstract. Identifying the suitable set of optimization options and determining nearly optimal values for the compiler parameter set in modern day compilers is a combinatorial problem. These values not only depend on the underlying target architecture and application source, but also on the optimization objective. Standard optimization options provide inferior solutions and also often specific to a particular optimization objective. Most common requirement of the current day systems is to optimize with multiple objectives, especially among average execution time, size and power. In this paper we apply Genetic Algorithm using Weighted Cost Function to obtain the best set of optimization options and optimal parameter set values for the multi-objective optimization of average execution time and code size. The effectiveness of this approach is demonstrated with the benchmark programs from SPEC 2006 benchmark suite. It is observed that the results obtained with parameter tuning and optimization option selection are better or equal to the results obtained with '-Ofast' option in terms of execution time and at the same time equal to '-Os' option in terms of code size.

Keywords: Compiler Optimization, Optimization Option Selection, Parameter Tuning, Genetic Algorithms, Multi-Objective Optimization.

1   Introduction

The quality requirements of the modern day systems are manifold. These systems need to execute the applications with lesser execution time, with minimal power intake and if possible with lesser memory. These requirements differ from the system to system. For example, main priorities of high performance computing and high-end embedded systems are execution time and energy. Similarly for low-end embedded applications like wireless sensor nodes 'motes', components of Internet of things (IOT) etc., prefer to optimize for energy and code-size. To meet these challenges compiler has to play a significant role and provide better solutions from multi-objective perspective. Modern compilers usually provide a vast spectrum of optimizations which intend to improve the code in number of aspects like execution time, compile time, code size, energy consumption etc. This wide variety of sophisticated optimizations includes local optimizations, global optimizations, inter-procedural optimizations, feedback directed optimizations, link-time optimizations etc. which can further be classified as

M.N. Murty et al. (Eds.): MIWAI 2014, LNAI 8875, pp. 23–34, 2014.

architecture dependent and architecture independent optimizations [8]. Anticipating the efficacy of these optimizations is not trivial. The effect of these optimizations depends on various factors like, underlying target architecture and system, application source, optimization objective etc. Present compilers generally provide some standard optimization options like -O0, -O1, -O2, -O3, -Os, -Ofast etc., each of them invokes a unique and predefined set of optimization options along with parameter values and intend to optimize either execution time and/or code size. Among these standard optimization options, -Ofast is meant to generate executables with best execution time and -Os generates executable with minimal footprint. -O2 balances both the aspects of objectives and generates executable which is reasonably faster in execution and good at code size as well. These pre-defined set of optimization options were decided by the compiler writers. However, due to the lack of underlying application source knowledge, these standard options are often not very effective. Study shows that standard optimization levels result in poor performance [4, 5, 6, 7, 9], and there is a necessity for more refined approaches.

Tuning the compiler settings and thereby turning on or off various compiler optimization settings results in maximal performance [1]. Sophisticated auto-tuning strategies for exploring the optimization sequences is considered as the one of the major sources of the unexploited performance improvements with existing compiler technology [2]. Optimization options with the recent compilers are plenty. Many of these optimization passes have hard-coded parameters set by the compiler writer, which may not produce the most optimal code. Due to the sheer number of optimizations available and the range of parameters that they could take, it becomes impossible to identify the best sequence by hand [3]. The search for suitable optimization sequences and optimization parameters that promise a positive effect on a single or multiple objective functions is not straightforward.

Current paper deals with the tuning of the parameter set and also optimization option selection with a multi-objective consideration. We have taken the average execution time and code size as the multiple objectives for our study. For this multi-objective experimentation, we are applying the Genetic Algorithm using weighted cost function with prior articulation of the preferences on objectives. The advantage of this approach is that it is fairly simple and fast compared to other Pareto front approaches and results in a unique single outcome [13, 14]. Objective of the current study is to obtain the best set of optimization options and parameter set values, which can yield a better result with respect to both the objectives, i.e., executable whose execution time is near to '-Ofast' and at the same time code size near to the one generated with '-Os'.

The rest of the paper is structured as following. Section 2, explores the compiler optimization space. Section 3 illustrates the multi-objective exploration framework, which includes the details of experimental setup, strategy followed to fine tune the parameter set and select the optimization options, fitness function and issues faced. Section 4 analyzes the results obtained and checks the effectiveness of parameter tuning and optimization option selection strategies. Finally in section 5, the results were summarized and a case is made for future research in this area.

2   Compiler Optimization Space

As discussed, modern day compilers, both commercial and open-source, are equipped with vast number of optimizations with complex mutual interactions among them. Effects of these optimizations with respect to various objectives such as execution time, code size or power are hardly predictable. For our study we consider the widely known and used open-source GCC compiler version 4.6.4 and its optimization space. This compiler does many optimization transformations in more than 200 passes on both GIMPLE and RTL. Some of these passes are architecture specific and also some of the passes are especially related to constant propagation and dead code elimination etc., are invoked multiple times.

Optimizations offered by this compiler are mainly of two types: those that apply to all architectures, controlled by -f options and those that are target-specific, controlled by -m options. There are around 176 optimization options of -f type supported by the GCC 4.6.4 compiler[10], which can be made on or off by the user. GCC implements the notion of optimization levels, which are umbrella options that automatically enable individual transformations. These optimization levels include O0 to O3 and Os which will enable or disable certain individual optimization options. -O0 is the default level which is meant for less compile time and better debug information. -O1 to -O3 levels will gradually increase their stress on execution time at the cost of increasing compilation time, increasing code size and decreasing debugging information. The main objective of the optimization level -Os is to reduce the code size. Table 1 lists the number of optimization options enabled or disabled with different optimization levels. However, there is no guarantee that all these optimization levels will perform well on different architectures for various applications. Not all options are controlled directly by a flag and also not all options are enabled even with -Ofast level, mainly because either the specific transformation is still relatively immature or because it only benefits few programs [11].

Table 1. List of enabled and disabled optimizations for different optimization levels

Optimization level	No of Enabled Optimizations	No of Disabled Optimizations
-O0	49	127
-O1	74	102
-O2	101	75
-O3	108	68
-Ofast	108	68
-Os	102	74

Apart from these optimization options, GCC provides around 120 parameters specific to the entire spectrum of optimizations like alignment related, branch & loop related, local & global optimizations, inter-procedural analysis based, register-allocation related, link time optimizations etc. These parameters correspond to various optimization pass implementations and were provided by the compiler writers, so that advanced users can fine-tune them. GCC provides an option of the form '–param name=value', to change these parameter values explicitly. The 'name' refers to the parameter name and the 'value' is the allowed value from its range. Each parameter takes a discrete value from its unique allowed range with default, minimum and maximum values as provided in the GCC internal documentation. For example, delayed branching optimization, which is mainly relevant in RISC machines can be invoked using the -fdelayed-branch option, is supported by parameters 'max-delay-slot-insn-search' & 'max-delay-slot-live-search' whose values can be fine-tuned by the user. Tuning of these parameters will have any impact only when the corresponding optimization options are invoked. Unlike the options, neither reordering nor repetition of these parameter settings will have any impact. We have considered 104 parameters for our study and rest of the parameters are not considered, as their values are fixed as per host processor architecture, or they are not relevant from the code size and execution time perspectives, or enough documentation is not available.

The literature survey shows many attempts by various researchers to tune the optimization options using various approaches including the statistical tuning [1], genetic algorithms based [7] and also machine learning based [6]. It is intuitive that the parameter set also plays an important role in achieving better performance. Fine tuning of the parameter set is not considered by earlier researchers. In our experimentation we study the impact of the parameter set on the specific objectives of code size and execution time and see whether the impact of parameter tuning is significant or not.

3   Multi-objective Exploration Framework

3.1   Experimental Setup

Intel Xeon E540 based 4 core system, each core operating at 2.66GHZ with 6MB of L1 cache and with Fedora release 17 having Linux Kernel 3.6.9 is used for the experimentation. This framework is developed using MATLAB and bash scripting. The benchmark cases are selected from CINT2006 (Integer Component of SPEC CPU2006) and CFP2006 (floating point component of SPEC CPU 2006) benchmark suites [12]. These are compute-intensive benchmark programs written in 'C' and details are as follows. The benchmark programs are 'bzip2', which handles compression, 'gobmk' is Artificial intelligence based GO game implementation, h264 format video compression algorithm implementation called 'h264ref', 'hmmer' for the gene sequence search, quantum computing related 'libquantum', combinatorial optimization based program 'mcf', perl programming language 'perlbench', artificial intelligence based chess program 'sjeng'. We have considered the GCC 4.6.4 compiler on X86 platform for this experimentation.

3.2   GA Based Optimization Option Selection and Parameter Set Tuning

Genetic Algorithm is an optimization and search technique based on the principles of genetics and natural selection. It simulates the natural evolutionary process of survival of the fittest and it evolves until an optimum is reached. For the current problem of obtaining the best optimization options and parameter values, we have used the coded values of compiler runtime parameters and/or optimization options are referred as genes. We have considered 104 compiler parameters p_1, $p_2...p_{104}$ and 105 optimization options O_1, $O_2...O_{105}$ provided by GCC 4.6.4 based C compiler on Intel X86 platform, for our study. Main criteria for choosing these parameters and optimization options are suitable with the targeted objectives, relevance to the target platform, availability of documentation, maturity level of options etc.

We have carried out the GA based Optimization option selection and parameter tuning independently and also together. The parameters can be supplied to the compiler using the command line switch of type '*--param name=value*', where name refers p_i and value is a discrete value taken from its corresponding range using the encoding function. The range of each variable is mapped with the gene length of 8bits using the encoding function. Optimization option selection information can be provided to the compiler using '*-f*' options, where '*-fno-optionName*' refers to the non-selection and '*-foptionName*' indicates selection of the respective optimization option. Chromosome for parameter tuning experiment will contain the string of genes of type '*--param name=value*' and for optimization option selection experiment it is of type '*-foptionName*'. For the problem of optimization option selection and parameter tuning, chromosome will contain the string of genes of both types.

Each chromosome represents possible solution to the problem and set of all such chromosomes will constitute the population. The chromosome will be applied to the cost function, which in turn invokes the compiler under study and provides the fitness value. Details of the cost function and fitness value are given in the next section. As the values that all parameters can take are only discrete values from their respective range, we have used the Binary Genetic Algorithm for experimentation. The initial population was selected randomly i.e. the initial values of all genes (parameters/ optimization options) in each chromosome (compilation sequence) of the initial population was taken randomly.

Table 2. Lists GA Properties and their values considered

GA Property	Value	GA Property	Value
Number of Generations	100	Selection Method	Weighted Random Pairing
Number of chromosomes in each Generation	16	Crossover Method	Single Point
Number of Genes in each Chromosome	104/166 /270	Crossover Probability	0.5
Gene Length	8 bit	Mutation rate	0.15

The evolutionary process involves cycles of generating and testing new offspring solution using one of two operations: crossover and mutation. Using crossover operation we combine part of one chromosome with part of another chromosome and create a new chromosome. Similarly mutation operator is applied to change the genes randomly with mutation rate probability. The GA based evolution is repeated for a fixed number of generations and the best chromosome is selected. Values of different GA parameters used are given in table 2.

3.3   Fitness Function

For the current study, we have considered multiple objectives viz., average execution time and code size. Often there is a trade-off between these objectives. For example, loop unrolling will reduce the execution time but increases the code size. Various approaches are provided in the literature to solve the multi-objective optimization problems using variants of Genetic Algorithms. They are mainly algorithms based on Pareto front (NSGA, VEGA and MOGA)[13] etc., which provide a set of optimal solutions. Other simple and most straight forward approach to multi-objective optimization is to obtain the single weighted cost function with prior articulation of the objective preferences. The problem with this method is to determine the appropriate values for weights w_n. This approach is not computationally intensive and results in a single best solution based on the assigned weights. We have employed this technique for our study and fixed the values of weights as '3' for code size and '7' for execution time in the total scale of 10. These are just representative numbers indicating the priority assigned to the respective objective and we want to give more priority to execution time than the code size. In order to obtain the fitness value of each chromosome, we follow the following steps. Firstly, the code size corresponding to the chromosome under study is obtained. For this, we invoke the compiler with current chromosome and apply the 'size' command on this generated executable. The sum of text and data section sizes was taken as the code size. Similarly, execution time of a chromosome is obtained by applying the 'time' command during the execution of the binaries, which is generated by the compiler after invoking with the corresponding chromosome. We repeat this process for a fixed number of times and obtain the average of these execution times. This process is to eliminate the impact of scheduling and context switch delays, I/O delays etc. Next, we normalize these code size and execution times with the corresponding values obtained using '-O2' option. Subsequently, weighted mean of the normalized code size and execution time with corresponding stated weights were computed and this weighted mean is considered as the fitness value of the present chromosome. As the objective under consideration is to minimize both the objectives i.e., code size and average execution time with its own weights, the Chromosome with low fitness value is desirable.

3.4   Issues Faced during GA Framework

Many problems/issues were faced during the execution of the GA framework for handling the compiler parameter tuning and optimization option selection problems. As part of the experimentation, the compiler was invoked with many optimization options which were rarely used and most of them were not enabled by any standard optimization options. Also, the many values assigned to several parameters were rarely used. Due to these reasons, compiler was subjected to many unexplored or least traversed control paths and also subjected to several corner cases, resulting to various compilation and execution problems. These problems may be rooted in various components of the compilation system including binutils (the open-source package which provides set of binary utilities like assembler, linker, disassembler, object copier etc.,), compiler, libraries and CRT (C runtime).

Further, these problems can be broadly classified as compile time and execution time problems. Some of the compilations issues observed are like hanging, Internal

Compiler Errors, floating point exceptions etc. These are due to several reasons, for example when the values of certain parameters like *'min-crossjump-insns'*, *'hot-bb-frequency-fraction'*, *'ira-max-loops-num'*, *'align-threshold'* etc., were equal to *'0'* then compiler emitted the Internal Compiler Error with various exceptions. Similarly, when the value of *'selsched-max-sched-times'* is taken as '0', the compiler hangs without any error messages. Several other issues were observed due to many reasons like optimization option incompatibilities, limitations of the compiler, usage of unsupported features etc.

Issues at execution stage may be manifested as too little execution time, too much execution time/hanging, core dump/segmentation fault/bus error etc. Possible reasons of this behavior is the usage of optimization options which are not mature and experimental in nature, options which are aggressive in nature and do not guarantee the correct result, incorrect code generation, issues in the target processor and its eco system, limitations of the target platform etc. When the execution time is less than one tenth of the executable time obtained while compiled with -Ofast and executed, then it is considered to be failure case. Both the compilation and execution errors were handled by generating another valid chromosome and replacing it with the problematic chromosome.

4   Analysis of the Results

As mentioned earlier, the experimentation was carried out with 8 benchmark programs taken from SPEC 2006. '-O2' was taken as the default standard optimization option being a normal practice. We have applied GA with weighted cost function of both the target objectives of code size and average execution time with pre-assigned weights of '3' for code size and '7' for execution time and conducted three kinds of experimentations. Firstly, tuning the selected 104 parameters, secondly optimization option selection for 105 optimization options and finally both the optimization option selection and also parameter tuning were applied together. The evolution process was carried out for 100 generations and it is observed that for many cases the cost is getting converged to an optimum level and local minima is reached. Cost is obtained by computing the weighted average of the normalized code size and execution times. Thus, the final results (list of optimization options and/or values of parameter set) obtained by the evolutionary process provides executable with lesser execution time and code size. We have observed that for most of the cases the size of the executable is better than or equal to the size obtained by the switch '-Os' and at the same time the execution time is better than or equal to the execution time with '-Ofast' switch. Fig.1 to Fig 8 demonstrates the code size and corresponding execution time taken at all the standard optimization option and also results obtained through the current experimentation. In these figures, O2PT refers to the results obtained subsequent to the parameter tuning with '-O2' as base standard optimization option. Similarly O2OS refers to the Optimization option selection with '-O2' as base optimization and finally O2PTOS refers to the results obtained as a result of both parameter tuning and optimization option selection with '-O2' as base optimization level.

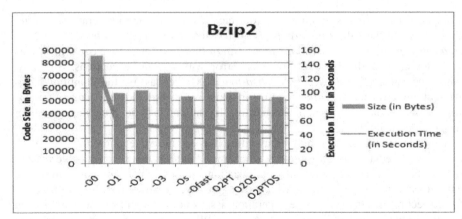

Fig. 1. Impact of Parameter Tuning and Optimization Option Selection for *'bzip2'* benchmark program

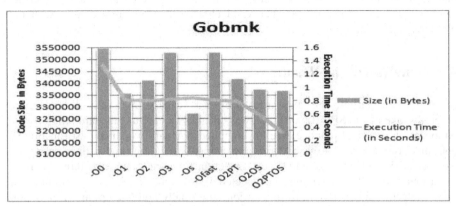

Fig. 2. Impact of Parameter Tuning and Optimization Option Selection for *'gobmk'* benchmark program

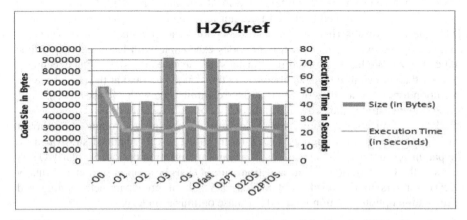

Fig. 3. Impact of Parameter Tuning and Optimization Option Selection for *'h264ref'* benchmark program

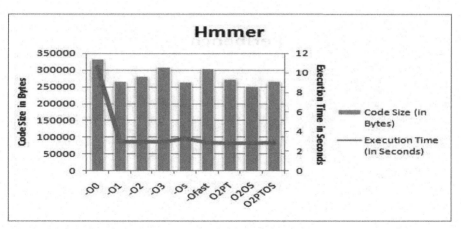

Fig. 4. Impact of Parameter Tuning and Optimization Option Selection for *'hmmer'* benchmark program

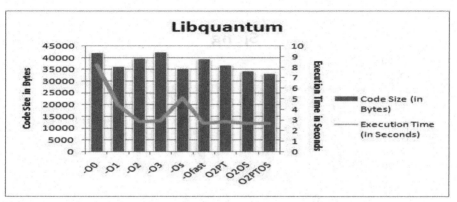

Fig. 5. Impact of Parameter Tuning and Optimization Option Selection for *'libquantum'* benchmark program

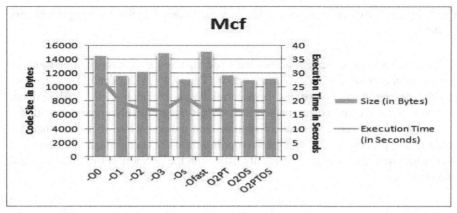

Fig. 6. Impact of Parameter Tuning and Optimization Option Selection for *'mcf'* benchmark program

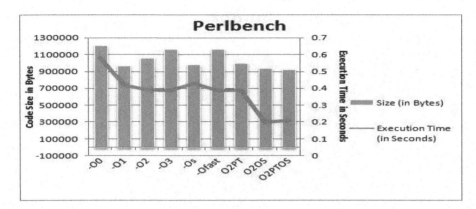

Fig. 7. Impact of Parameter Tuning and Optimization Option Selection for *'perlbench'* benchmark program

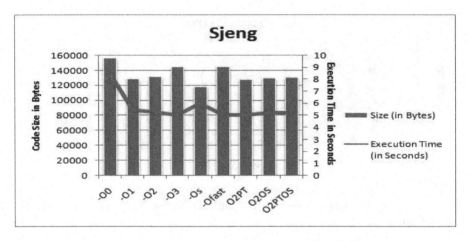

Fig. 8. Impact of Parameter Tuning and Optimization Option Selection for *'sjeng'* benchmark program

The correlation coefficients across the code size and execution time at standard optimization options and also at tuning options i.e., O2PT, O2OS and O2PTOS were calculated. Figure 9 demonstrates these correlation coefficients. It is clearly evident that both code size and corresponding execution time are highly correlated with the tuning options compared to standard optimization options. Thus, it is clear that GA based tuning process for Multi-objective optimization selects optimization settings in such a way that, they yield better results from both the perspectives of code size and execution time.

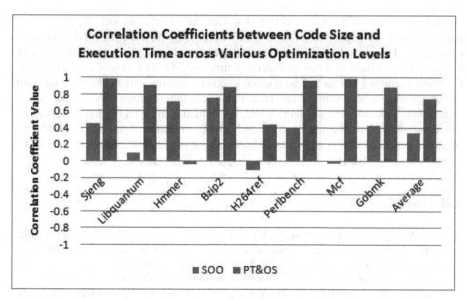

Fig. 9. Correlation coefficients between Code Size and Execution Time across Standard Optimization Options and Tuning Options

Observations: It is observed that, both the optimization objectives i.e., code size and average execution time is positively correlated. All the eight benchmark programs have shown positive correlation with average correlation value of 0.48. However, it is also observed that for almost all the programs with '-Os' option, meant for code size optimization, execution time is relatively higher. Similarly for almost all the programs, it is observed that with -Ofast optimization, which is meant to optimize the execution time, the code sizes are relatively higher. At the same time the results obtained with Parameter Tuning and Optimization Option Selection have demonstrated much better results both in terms of code size and execution time. Seven out of eight benchmark programs delivered best execution timings with either Parameter tuning, Optimization Option Selection or both. For code size, four programs demonstrated best results. It is also observed that, while the parameter tuning and optimization option selection has positive impact over the standard optimization options, parameter tuning over optimization selection has marginal impact in four benchmark programs, reason for the same needs to be further investigated. Probably, allowing the GA process to evolve for more number of generations, especially considering the gene length (105 + 104 = 209) and tuning the various other GA parameters will resolve the issue.

5   Summary and Future Work

This study brings out the fact that by applying parameter tuning and optimization option selection strategies, we can obtain much better results which are good from both the optimization objectives of code size and execution time. With the current experimentation results, it is evident that Genetic algorithm with weighted cost function approach yields single and much better results from the multi-objective perspective. These results are much better than the results of '-O2', and simultaneously comparable to '-Ofast' in terms of execution time and '-Os' in terms of code size. This fine-tuning strategy can further be applied to other multi objectives like code size and power, power & execution time etc., based on the requirements of the target application domain. Fine-tuning strategies which can yield global optimum, techniques to reduce the tuning time and also optimization framework which can employ tuning and option selection technologies in production compilers etc., needs to be further investigated.

References

1. Haneda, M., Knijnenburg, P.M.W., Wijshoff, H.A.G.: Automatic Selection of Compiler Options using Non-Parametric Inferential statistics. In: 14th International Conference on Parallel Architectures and Compilation Techniques (PACT 2005) (2005)
2. Adve, V.: The Next Generation of Compilers. In: Proc. of CGO (2009)
3. Duranton, M., Black-Schaffer, D., Yehia, S., De Bosschere, K.: Computing Systems: Research Challenges Ahead The HiPEAC Vision 2011/2012
4. Kulkarni, P.A., Hines, S.R., Whalley, D.B., et al.: Fast and Efficient Searches for Effective Optimization-phase Sequences. Transactions on Architecture and Code Optimization (2005)
5. Leather, H., O'Boyle, M., Worton, B.: Raced Profiles: Efficient Selection of Competing Compiler Optimizations. In: Proc. of LCTES (2009)
6. Agakov, F., Bonilla, E., Cavazos, J., et al.: Using Machine Learning to Focus Iterative Optimization. In: Proc. of CGO (2006)
7. Cooper, K.D., Schielke, P.J., Subramanian, D.: Optimizing for Reduced Code Space using Genetic Algorithms. SIGPLAN Not. 34(7) (1999)
8. Khedkar, U., Govindrajan, R.: Compiler Analysis and Optimizations: What is New? In: Proc. of Hipc (2003)
9. Beszédes, Á., Gergely, T., Gyimóthy, T., Lóki, G., Vidács, L.: Optimizing for Space: Measurements and Possibilities for Improvement. In: Proc. of GCC Developers Summit (2003)
10. GCC, the GNU Compiler Collection - online documentation, http://gcc.gnu. org/onlinedocs/
11. Novillo, D.: Performance Tuning with GCC. Red Hat Magazine (September 2005)
12. SPEC-Standard Performance Evaluation Corporation, http://www.spec.org/ cpu2006
13. Haupt, R.L., Haupt, S.E.: Practical Genetic Algorithms. Wiley Interscience (2004)
14. Timothy Marler, R., Arora, J.S.: The weighted sum method for multi-objective optimization: New insights. Journal on Structural and Multidisciplinary Optimization (2009), doi:10.1007/s00158-009-0460-7

Association Rule Mining via Evolutionary Multi-objective Optimization

Pradeep Ganghishetti[1,2] and Ravi Vadlamani[1,*]

[1] Centre of Excellence in CRM & Analytics,
Institute for Development and Research in Banking Technology,
Hyderabad-500 057, A.P, India
[2] SCIS, University of Hyderabad, Hyderabad-500046, A.P, India
pradeepghyd@gmail.com, rav_padma@yahoo.com

Abstract. In this paper, we formulated association rule mining as a combinatorial, multi-objective global optimization problem by considering measures such as support, confidence, coverage, comprehensibility, leverage, interestingness, lift and conviction. Here, we developed three evolutionary miners viz., Multi-objective Binary Particle Swarm Optimization based association rule miner (MO-BPSO), a hybridized Multi-objective Binary Firefly Optimization and Threshold Accepting based association rule miner (MO-BFFOTA), hybridized Multi-objective Binary Particle Swarm Optimization and Threshold Accepting based association rule miner (MO-BPSOTA) and applied them on various datasets and conclude that MO-BPSO-TA outperforms all others .

Keywords: Evolutionary Multi-objective Optimization, Association rules, Crisp Product Operator, Combinatorial Global Optimization, Quality Measures.

1 Introduction

Association rule mining is a very important data mining task used to extract important correlations among the products from transactional databases [1]. An association Rule is of the form A→B, where A and B represent item sets (I) or products, and an item set includes all possible items{i_1, i_2, . . ., i_m} in a transactional database. The a priori algorithm [2] works in two phases- generation of frequent item-sets and rule generation. Later, Han et al. [3] proposed FP-Growth algorithm that generates a F-List and a tree followed by mining of rules. Association rules have been extracted by leveraging evolutionary computational techniques. Saggar et al. [4] optimizes association rules extracted by a priori via Genetic Algorithm (GA). Anandhavalli et al. [5] also uses GA based approach for extracting association rules with negative attributes. Multi-objective GA based approaches have also been suggested [6-8]. Kaya et al.[9] extracted partial optimized fuzzy association rules. Then, particle swarm optimization (PSO) was used for association rule mining [10,11]. Nandhini et al. [12] developed domain ontology and PSO based association rule mining algorithm. Alatas et al. [13]

---

* Corresponding author.

M.N. Murty et al. (Eds.): MIWAI 2014, LNAI 8875, pp. 35–46, 2014.
© Springer International Publishing Switzerland 2014

devised pareto-based multi-objective differential evolution (DE) for extracting association rules. Kuo et al. [14] used PSO to objectively and quickly determine user-defined parameters. Menéndez et al [15-17] proposed Multi-objective Genetic Algorithm (MOGGC) for spectral Clustering problem, feature selection followed by clustering and image segmentation. Most recently, Sarath and Ravi [18] develooped binary PSO to extract association rules from datasets without having to specify the minimum support and confidence upfront. Naveen et al. [19] devised firefly optimization based rule extractor for classification Maheshkumaret al [20] proposed PSO-TA hybridized algorithm for solving unconstrained continuous optimization.

2   Various Optimization Based Techniques

2.1   Firefly Optimization Algorithm (FFO)

The Firefly algorithm [21] is a population based global optimization technique inspired by the natural behavior of fireflies. Each firefly moves to more brighter/attractive firefly wherein firefly's brightness is characterized by its objective function. The attractiveness of a firefly is directly proportional to its brightness and decreases as the distance from the other firefly increases. Here each firefly represents a solution in the optimization parlance. The attractiveness (β) is a monotonically decreasing function of the distance r between any two fireflies.

$$\beta = \beta_0 e^{-\gamma r^2} \tag{1}$$

The traversal of a firefly towards other brighter fireflies is given by:

$$x_i(t+1) = x_i(t) + \beta_0 e^{-\gamma r^2}(x_i - x_j) + \alpha(\text{rand}() - 0.5) \tag{2}$$

The algorithm of the firefly optimization is depicted in the box as follows:

```
Algorithm:
•Define Objective function f(x),        x=(x₁,....xₐ)ᵀ, β₀, γ
•Generate initial population of fireflies X ᵢ (i=1,2,...,n)
• Light intensity Lᵢ at xᵢ is determined by f(xᵢ)
•While(t<MaxGeneration)
•for i=1:n all n fireflies
•        for j=1:n all n fireflies
•                if(Lⱼ>Lᵢ), Move firefly i towards j in d-dimensions as in eqn 1 and 2 end if
•        Evaluate new solutions & update light  intensity
•        end for j
•end for i
•Rank the fireflies and find the current best
•end while
```

where t denotes iteration number, rand() indicates a random number between 0 and 1, β_0 denotes the attractiveness constant and γ is the light absorption coefficient. After every iteration, randomness parameter (α) is reduced by a constant Delta (Δ). The distance between two fireflies i and j at positions x_i and x_j can be defined as follows:

$$r_{ij} = \|x_i - x_j\| = \sqrt{\sum_{k=1}^{d}(x_{ik} - x_{jk})^2} \tag{3}$$

2.2   Threshold Acceptance Algorithm (TA)

Threshold Accepting algorithm, proposed by Deuck and Sheuer [22] is a point that is not much worse than the current one. A candidate solution Cand_Soln[i] in the neighborhood following rule scheme presented in Table 1 is generated as follows:

```
Algorithm:
•Consider weakest particle/firefly of MO-BPSO/MO-BFFO as the initial solution and let its fitness
be fᵢ.
•for each of the global iterations
•        While(innerIterationNo<MaxInnerIterations or Δ1 > thresh)
•        Generate a candidate solution in the neighborhood and let its fitness be fⱼ.
•        Compute deterioration Δ1=fᵢ-fⱼ
•        If Δ1 < thresh, set fᵢ = fⱼ                //To accept/reject the solution
•        If thresh <thrtol, set Δ2 = (new - old) / old    //for execution of ample no. of global iters
•        Report current solution as the optimal one if abs(Δ2) <acc and exit if end of global iters
•        Else
•        old=new
```

```
for(j=0;j<2×n;j++)                          //generation of a candidate solution
    If(rand()<bias)        Cand_Soln[i][j]=1
    else        Cand_Soln[i][j]=0
```

Table 1. Rule representation

I_1		I_2		I_3		...	I_N	
V_{11}	V_{12}	V_{21}	V_{22}	V_{31}	V_{32}	...	V_{N1}	V_{N2}

Likewise, we generate rules of varied lengths (like 2,3,4,..). If there are 25 features in the dataset, the bias value of 0.1 is found to be suitable after several computations. The bias to be fixed is inversely proportional to the number of features in the dataset ($bias_1 \times featuresCount_1 = bias_2 \times featuresCount_2$).

2.3   Binary Firefly Optimization (MO-BFFO)

We developed a combinatorial version of MO-BFFO to solve our optimization problem. Here, each firefly's positional value in each dimension can be either 0 or 1 only. This value of x_i is determined probabilistically depending on the changing rate of real value of x_i. For binary version of firefly optimization, the firefly's positions are updated as follows:

$$x_{i+1} = x_i + \beta_0 e^{-\gamma r^2}(x_i - x_j) + \alpha(rand - 0.5)$$
```
If (rand () < S (xᵢ₊₁))
Then    xᵢ₊₁=1
Else    xᵢ₊₁=0
```

Where S is the sigmoid function, rand () is the uniform random number between (0, 1), β_0 is the attractiveness constant, γ is the absorption coefficient and r is the Euclidean distance between the two fireflies i and j. The sigmoid function used in our algorithms is as follows: $S(x) = \frac{1}{1+e^{-x}}$. Binary PSO was developed in [18] as a binary version of PSO [23]. For more details the reader is referred to [18].

3   Multi-objective Association Rule Miners

3.1   Preprocessing

The first step involves transformation of data into binary encoded format where each record is stored in terms of 0's and 1's [24] as in fig 1. This step is necessary for faster database scanning and faster calculation of various measures. Let there be transactions T_1-T_5 with five different items I_1-I_5. For instance, in transaction T_5, the values of cells I_1 and I_5 are both "1's" whereas cells I_2, I_3, I_4 are "0's" indicating items I_1 and I_5 are only purchased. The second step of preprocessing is feature selection used when the number of features is above 50. Here, we discard all the features whose item support is less than α, where α is very small user defined value. This helps in obviating the unnecessary computations and removing extremely rare rules.

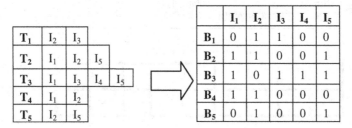

Fig. 1. Binary Transformation of the original dataset

3.2   Rule Representation

In this paper, we followed Michigan approach [25], where each chromosome represents a separate rule. Let there be N number of items in the dataset. Each item is represented by two bits and each bit can take the values either 1/0. The value of 1 in the first bit indicates the item is present otherwise absent in the rule. The second bit signifies whether the item is included in the antecedent (i.e., 1) or consequent (i.e., 0).

11-item present in antecedent   10-item present in consequent   00/01-item absent in the rule

3.3   Quality Measures

In most cases, the quality of association rules cannot be judged by considering only support and confidence but there are other measures as follows. Let a rule be represented as A→B where A is the antecedent and B is called the consequent.

1. Support
Support is defined as the percentage or fraction of transactions in the database that contain both antecedent as well as the consequent parts.

2. Confidence:

Confidence indicates how **reliable or relevant** a given rule is. Confidence is defined as the probability of occurring the rule's consequent under the condition that the transactions also contain the antecedent.

3. Interestingness:

A rule is said to be interesting when the individual support count values are greater than the collective support (A→B) values.

$$Interestingness = \frac{Support(AUB)}{Support(A)} \times \frac{Support(AUB)}{Support(B)} \times \left(1 - \frac{Support(AUB)}{Support(D)}\right)$$

4. Comprehensibility:

In association rule mining, if the number of conditions involved in the antecedent part is less than the one in the consequent part, the rule is more comprehensible.

Comprehensibility=log(1+|c|/log(s+|A U C|))

5. Lift:

$$Lift(A \to B) = \frac{Confidence(A \to B)}{Support(B)} = \frac{Support(A) \times Support(B)}{Support(B)}$$

The lift value is a measure of **importance** of a rule (originally called interest)

6. Leverage:

The rationale in a sales setting is to find how many more units (items X and Y together) are sold than expected from the independent sells and shows the impact of ARM.

$$Leverage(A \to B) = Support(A \to B) - Support(A) \times Support(B)$$

7. Conviction:

$$Conviction(A \to B) = \frac{(1 - Support(B))}{(1 - Confidence(A \to B))} = \frac{P(A)P(!B)}{P(Aand\,!B)}$$

Conviction compares the probability that Antecedent appears without Consequents if they were dependent with the actual frequency of the appearance of Antecedents without Consequents.

8. Coverage

$$Coverage = Support(A)$$

Coverage measures how often a rule X -> Y is applicable in a database. Objective function

The choice of fitness/objective function is helpful in assessing the importance of each individual firefly in the population set. Here, we assign equal weightage to all the above defined measures and hence use the product of all of them to compute the objective function as follows:

$$Fitness = \prod_{i=1}^{8} Measure\_Value(i) \tag{4}$$

3.4   Steps Involved in MO-BPSO

In MO-BPSO, we proposed the fitness our proposed model, we need to run the algorithm M times in order to get all the top M rules in the database. In every run, we store the positions and movements of all the fireflies in the population for each of the iterations. After all the runs, the records are ranked and duplicate records are removed. We pick the top M rules as per the objective function. The main idea behind running the algorithm M times is as follows. Whenever evolutionary algorithms are employed to classification or association rule mining problem, the user gets different

Table 2. Parameters chosen for MO-BPSO Rule Miner

Dataset	N	Inertia	C1	C2	Max Iterations
Book	30	0.8	2	2	50
Food	30	0.8	2	2	50
Grocery	30	0.8	2	2	100
Bank	50	0.8	2	2	50
Click stream	30	100.8	2	2	50
Bakery	30	0.8	2	2	50

rules in every run. Therefore, running the MO-BPSO many times and collating the rules obtained indeed becomes an ensembling scheme for extracting rules. The parameters used in MO-BPSO are presented in Table 2.

3.5   Steps Involved in MO-BFFO-TA

Initialize each firefly randomly with either 0 or 1 such that fitness values are greater than 0.0001

Repeat the following steps until Max. Iterations
- Calculate the objective function for all fireflies using eqn.4.
- Replace the weakest Firefly in the population by the solution yielded by Threshold Accepting Algorithm, which is invoked probabilistically.
- For each firefly x_i, we compare its light intensity L_i with the light intensity L_j of every other firefly.
- If $L_i<L_j$, then we move firefly x_i towards x_j in n $_d$imensions using eqns. (2) and (3) and apply sigmoid function to transform real values to binary positions. We do not need to use the bias component in this approach.
- Compute the new value of the objective function for each firefly x_i and update its light intensity.

In MO-BFFO-TA algorithm, we capture the positions and movements of all the fireflies in each of the iterations of MO-BFFO-TA into a database. Then records are ranked and duplicate records are removed. We pick the top M rules as per the objective function. The parameters of MO-BFFO-TA are presented in Table 3.

The following box briefly describes the algorithm:

```
RS = /* initially, Rule Set is empty */
WHILE Iteration Number < Maximum Iterations
        IF rand()<Prob_TA
            Call the Threshold Accepting algorithm to replace theweakest firefly.
            Add all the candidate rules generated in the TA based search to the RS
            Replace the weakest firefly in the population
        ELSE
            Continue MO-BFFO ARM
            Add the rules of all the fireflies in the population of MO-BFFO to the RS in the current
iteration
Rank all the Rules in RS according to objective function
Discard all the duplicate rules
```

Table 3. Parameters chosen for MO-BFFO-TA Rule Miner

Dataset	N	β_0	γ	Δ	α	MaxIterations	Prob_TA	eps	acc	ThreshTol	Thresh	GI	IO
Book	10	2	2.5	1	0.5	50	0.2	0.01	0.5	0.00018	0.0002	100	25
Food	10	2	2.5	1	0.5	50	0.3	0.01	0.5	0.00018	0.0002	100	25
Grocery	10	2	2.5	1	0.5	50	0.3	0.01	0.5	0.00018	0.0002	100	25
Bank	10	2	2.5	1	0.5	150	0.2	0.01	0.5	1.8E-07	2E-07	200	25
Click stream	10	2	2.5	1	0.5	50	0.3	0.01	0.5	0.000018	0.00002	200	25
Bakery	10	2	2.5	1	0.5	200	0.8	0.01	0.5	1.8E-06	0.000002	200	25

3.6   Steps Involved in MO-BPSO-TA

Initialize each particle randomly with either 0 or 1 such that fitness values are greater than 0.0001

Repeat the following steps until Max. Iterations
- Calculate the fitness for all particles using eqn. 4.
- Replace the weakest particle in the population by the solution yielded by Threshold Accepting Algorithm, which is invoked probabilistically, one's or twice in hundred iterations
- Update the local best and global best values.
- Update particles position as in the usual PSO, wherein sigmoid function is applied to transform real values to binary positions.

In MO-BPSO-TA algorithm, we capture all the particles positions and movements in each of the iterations of MO-BPSO and TA into a database. Then the solutions are ranked and duplicate ones are removed. We pick the top M rules as per the objective function. The parameters of MO-BPSO-TA are presented in Table 4. The following box briefly describes the algorithm:

```
RS = /* initially, Rule Set is empty */
WHILE Iteration Number < Maximum Iterations
        IF rand()<Prob_TA
                Call the Threshold Accepting algorithm to replace the weakest firefly.
                Add all the candidate rules generated in the TA based search to the RS
                Replace the weakest firefly in the population
        ELSE
                Continue MO-BPSO ARM
                Add the rules of all the particles in the population of MO-BPSO to the RS in the current
iteration
Rank all the Rules in RS according to fitness function
Discard all the duplicate rules
```

Table 4. Parameters chosen for MO-BPSO-TA Rule Miner

Dataset	n	Inertia	C1	C2	MaxIterations	Prob_TA	eps	Acc	Thresh Tol	Thresh	GI	II
Book	30	0.8	2	2	50	0.1	0.01	0.5	0.000018	0.00002	50	5
Food	30	0.8	2	2	50	0.3	0.01	0.5	0.00018	0.0002	50	5
Grocery	30	0.8	2	2	100	0.1	0.01	0.5	0.00018	0.0002	50	5
Bank	50	0.8	2	2	50	0.7	0.01	0.5	1.8E-06	0.000002	50	5
Click stream	30	100.8	2	2	50	0.1	0.01	0.5	8E-09	1E-08	50	5
Bakery	30	0.8	2	2	50	0.1	0.01	0.5	0.000018	0.00002	300	25

GI-Global Iterations; II-Inner Iterations

4   Results and Discussion

We conducted the experiments on a system with Microsoft Windows 7 64-bit Operating System, Intel Core i5 processor, clock speed of 2.53GHz and 4 GB RAM. The proposed algorithms were developed using Java Standard Edition (JDK 1.7) with Eclipse IDE. In this paper, we considered six datasets for demonstrating the effectiveness. The first dataset in our study is Books dataset taken from XLMINER tool (www.solver.com/xlminer-data-mining). It includes 10 book types and 2000 customer records. Another dataset considered here is Food dataset taken from IBM SPSS Modeler (www.ibm.com/software/analytics/spss) tool. This dataset contains 11 types of various food items and 1000 transactional records. The third dataset is the grocery dataset taken from SAS enterprise tool (http://www.sas.com/technologies/ analyt-ics/datamining/miner). This dataset contains 20 grocery products and 1001 transactions. We also analyzed real world dataset from XYZ bank, a commercial bank. It is a sparse dataset consisting of 12191 customers' transactional records and 134 different product and service offerings. The fifth dataset is the Bakery dataset (https://wiki.csc.calpoly.edu/datasets/wiki/ExtendedBakery) which has a list of 40 pastry items and 10 coffee drinks with 1000 transactions. The last dataset is the Anonymous Web Dataset (http://archive.ics.uci.edu/ml/datasets/Anonymous+Microsoft+Web+Data). The data records include the use of www.microsoft.com by 37711 anonymous, random-ly-selected users where the dataset contains the lists all the areas of the web site (Vroots) that user visited in a one week timeframe. There are 294 features of website links in this dataset. The results (see Tables 5through 10) obtained are discussed as follows. Table 11 presents the computational times for all the three algorithms.

Table 5. Results of Books dataset

	Supp.	Conf.	Cov.	Comp.	Lev.	Int.	Lift	Conv.	Fitness
MO-BPSO	12.58	76.33	18.5	62.25	5.58	34.6	7.39	6.9	0.001506
MO-BPSO-TA	16.44	68.81	24.36	61.3	6.72	27.94	3.16	2.11	0.001096
MO-BFFO-TA	17	68.45	25.32	64.71	6.35	26.13	2.79	2.01	0.000997

Table 6. Results of Food dataset

	Supp.	Conf.	Cov.	Comp.	Lev.	Int.	Lift	Conv.	Fitness
MO-BPSO	13.25	71.15	20.64	55.16	7.07	25.57	2.37	3.7	0.002033
MO-BPSO-TA	12.19	71.41	19.37	54.47	6.38	23.08	2.37	3.78	0.00156
MO-BFFO-TA	9.5	50.88	18.4	60.64	4.63	16.12	2.66	1.5	0.000723

Table 7. Results of Grocery dataset

	Supp.	Conf.	Cov.	Comp.	Lev.	Int.	Lift	Conv.	Fitness
MO-BPSO	14.64	81.12	18.81	65.12	9.97	45.91	4.03	8.72	0.012162
MO-BPSO-TA	21.84	77.93	29.54	63.74	9.25	36.16	2.5	2.7	0.004953
MO-BFFO-TA	12.56	29.98	4.28	69.56	2.7	13.4	1.41	1.1	0.000265

Table 8. Results of Bank dataset

	Supp.	Conf.	Cov.	Comp.	Lev.	Int.	Lift	Conv.	Fitness
MO-BPSO	2.04	54.25	4.54	65.32	1.77	20.95	12.93	3.01	7.52E-05
MO-BPSO-TA	2.15	53.79	4.35	67.94	1.84	18.41	9.79	2.49	7.26E-05
MO-BFFO-TA	1.11	28.62	4.85	64.71	0.69	6.93	4.07	1.65	5.76E-05

Table 9. Results of Bakery dataset

	Supp.	Conf.	Cov.	Comp.	Lev.	Int.	Lift	Conv.	Fitness
MO-BPSO	4.21	56.1	7.87	68.25	3.66	35.2	9.45	4.47	0.000408
MO-BPSO-TA	4.36	51.63	8.41	69.56	3.75	33.75	8.69	1.95	0.000254
MO-BFFO-TA	3.63	47.06	7.67	66.32	3.1	25.2	7.52	1.78	9.92E-05

Table 10. Results of Click Stream dataset

	Supp.	Conf.	Cov.	Comp.	Lev.	Int.	Lift	Conv.	Fitness
MO-BPSO	4.99	60.54	8.93	66.84	3.37	23.02	6.86	3.24	0.000179
MO-BPSO-TA	5.57	53.88	11.09	65.01	3.48	20.54	5.24	2.11	0.000134
MO-BFFO-TA	2.09	23	13.76	69.56	0.95	5.57	2.78	1.27	1.97E-05

Table 11. Computational times for the algorithms

Data Set	MO-BFFO-TA	MO-BPSO-TA	MO-BPSO
BOOKS	3.205 s	2.22 s	20.595 s
FOOD	1.36 s	1.65 s	11.26 s
GROCERY	42.22 s	4.15 s	24.47 s
XYZ Bank	51.81 s	110.49 s	923.10 s
Clickstream	92.02 s	61.51 s	1567.97 s
Bakery	49.79 s	34.433 s	311.55 s

A) Books Dataset

On Books dataset, we applied MO-MO-BPSO association rule miner with the following parameter settings. We have fixed number of particles as 30, inertia as 0.8, constants c_1 and c_2 as 2 and no of iterations as 50. Later, we applied Hybridized MO-BPSO-TA algorithm. We have chosen same MO-BPSO parameters. Here for TA, we chose inner iterations as 50, outer iterations as 5, accuracy as 0.5, epsilon as 0.1 and TA is called with a probability of 10% of number of iterations. Later, we applied Hybridized MO-BFFO-TA based Association rule miner. We chose the number of fireflies as 10, attractiveness constant (β_0) as 2, gamma (γ) as 2.5 and alpha (α) as 0.5, and TA is called with a probability of 20% of total number of iterations. Here, i.e., with MO-BPSO based association rule miner, it extracted rules with higher values of confidence, Interestingness, Lift, Conviction. MO-BPSO-TA extracted rules with higher levels of leverage. MO-BFFOTA produced higher support, coverage, comprehensibility. We conclude that MO-BPSO outperforms all others, as it extracted rules with many higher measures of strength. The MO-BPSO-TA produced fitness values which are near to that of MO-BPSO. But, when time complexity is critical, MO-BPSOTA is preferred. MO-BFFO-TA is found to produce inferior fitness values.

B) Food Dataset

For food dataset, all the parameters for MO-BPSO, MO-BPSOTA, and MO-BFFOTA are the same as for books dataset. However, the probability of calling TA is increased to 30% for MO-BPSO-TA and MO-BFFO-TA algorithms for increasing the efficiency of the algorithm as useful rules are in the limited neighborhood. MO-BPSO extracted rules with higher support, coverage, leverage and interestingness. We observed that MO-BPSO-TA produced higher confidence, conviction values, while MO-BFFO-TA produced rules with higher comprehensibility, lifts values. Similar to Books dataset, MO-BPSO outperformed all other operators on food dataset.

C) Grocery Dataset

Here too, the parameters for MO-BPSO, MO-BPSOTA, and MO-BFFOTA are same as for books dataset except that the maximum iterations for MO-BPSO is increased to 100 due to nature of the dataset. MO-BPSO produced rules with higher confidence, comprehensibility, leverage, Interestingness, lift and conviction values. However, MO-BPSOTA extracted rules with higher support and coverage values. MO-BFFO-TA is found be inferior for all the measures. Similar to Books dataset, MO-BPSO outperformed all others on grocery dataset.

D) Bank Dataset

For Bank dataset, all the parameters for MO-BPSO are same as for books dataset except that the number of particles is increased to 50 due to data sparseness. In case of MO-BPSO-TA, the number of particles is increased to 50 and the TA is called with a probability of 70% of total number of iterations. The parameters for MO-BFFO-TA algorithm are the same as for Books dataset except that the number of iterations is increased to 150. MO-BFFO-TA extracted rules with high coverage, while MO-BPSO yielded rules with higher confidence, Interestingness, lift and conviction values. MO-BPSO-TA extracted rules with high support, comprehensibility and leverage. Again, MO-BFFOTA is found be inferior. Similar to Books dataset, MO-BPSO outperformed all others.

E) Bakery Dataset

The parameters for MO-BPSO are same as for books dataset, while that of MO-BPSO-TA are same as that of books dataset except that the max inner and outer iterations of TA are fixed at 300 and 25 respectively. MO-BPSO extracted rules with higher confidence, interestingness, lift, conviction values. However, MO-BPSO-TA, extracted rules with high support, coverage, comprehensibility, leverage. Again, MO-BFFO-TA was found to be inferior. Similar to Books dataset, MO-BPSO outperformed all other algorithms.

F) Clickstream Dataset

The parameters for MO-BPSO, MO-BPSO-TA are same as that of books dataset except that the inertia for MO-BPSO is increased to 100.8. The parameters for MO-BFFO-TA are the same as that of food dataset. MO-BPSO extracted rules with higher confidence, interestingness, lift, conviction. MO-BPSO-TA extracted rules with higher support, leverage. MO-BFFOTA extracted rules with high coverage, comprehensibility. Here too, MO-BPSO outperformed all others.

5 Conclusion

We proposed three techniques viz., Multi-objective Binary Particle Swarm Optimization (MO-BPSO), a Multi-objective Binary Firefly optimization and Threshold Accepting (MO-BFFO-TA) and a Multi-objective Binary Particle Swarm optimization and Threshold Accepting (MO-BPSO-TA) to extract association rules from databases by optimizing several rule quality measures objectives. The advantage of proposed methods is that of the user need not specify minimum support and confidence. The MO-BFFO-TA and MO-BPSO-TA Rule Miners could generate all the top 10 rules in just a single run. This is a very significant improvement over MO-BPSO. Overall, MO-BFFO-TA and MO-BPSO-TA also consumed less time compared to a priori, FP-Growth. Further, these algorithms do not generate redundant rules.

References

1. Agrawal, R., Imielinski, T., Swami, A.: Mining association rules between sets of items in large databases. In: Proceedings of the 1993 ACM SIGMOD Conference on Management of Data, Washington, D.C., USA, pp. 207–216 (1993)
2. Agrawal, R., Srikant, R.: Fast algorithms for mining association rules. In: Proceedings of the 20th Int. Conference on VLDB Conference, Santiago, Chile, pp. 487–499 (1994)
3. Han, J., Pei, H., Yin, Y.: Mining frequent patterns without candidate generation. In: Conference on the Management of Data, SIGMOD 2000, Dallas, pp. 1–12. ACM Press (2000)
4. Saggar, M., Agrawal, A.K., Lad, A.: Optimization of association rule mining using improved genetic algorithms. In: Proceeding of the IEEE International Conference on Systems Man and Cybernetics, vol. 4, pp. 3725–3729 (2004)
5. Anandhavalli, M., Kumar, S.S., Kumar, A., Ghose, M.K.: Optimized association rule mining using genetic algorithm. In: Advances in Information Mining, pp. 1–4 (2009)
6. Waiswa, P.P.W., Baryamureeba, V.: Extraction of interesting association Rules using genetic algorithms. Int. Journal of Computing and ICT Research 2(1), 26–33 (2008)
7. Ghosh, A., Nath, B.: Multi-Objective rule mining using genetic algorithms. Information Sciences 163, 123–133 (2004)
8. Minaei-Bidgoli, B., Barmaki, R., Nasir, M.: Mining numerical association rules via multi-objective genetic algorithms. Information Sciences 233, 15–24 (2013)
9. Kaya, M., Alhajj, R.: Mining optimized fuzzy association rules using multi-objective genetic algorithm. In: 8th IEEE International Conference on Intelligent Engineering Systems, Cluj-Napoca, Romania, pp. 38–43 (2004)
10. Gupta, M.: Application of weighted particle swarm optimization in association rule mining. International Journal of Computer Science and Informatics 1 (2012) 2231–5292
11. Asadi, A., Afzali, M., Shojaei, A., Sulaimani, S.: New binary PSO based method for finding best thresholds in association rule mining. Applied Soft Computing, 260–264 (2012)
12. Nandhini, M., Janani, M., Sivanandham, S.N.: Association rule mining using swarm intelligence and domain ontology. In: IEEE International Conference on Recent Trends in Information Technology (ICRTIT), Coimbatore, pp. 537–541 (2012)
13. Alatas, B., Akin, E., Karci, A.: MODENAR: Multi-objective differential evolution algorithm for mining numeric association rules. Applied Soft Computing, 646–656 (2008)
14. Kuo, R.J., Chao, C.M., Chiu, Y.T.: Application of particle swarm optimization to association rule mining. Applied Soft Computing 11, 326–336 (2011)

15. Menéndez, H.D., Barrero, D.F., Camacho, D.: A multi-objective genetic graph-based clustering algorithm with memory optimization. In: IEEE Congress on Evolutionary Computation (CEC), pp. 3174–3181 (2013)
16. Menéndez, H., Bello-Orgaz, G., Camacho, D.: Features selection from high-dimensional web data using clustering analysis. In: Proceedings of the 2nd International Conference on Web Intelligence, Mining and Semantics, pp. 20:1–20:9. ACM, New York (2012)
17. Menéndez, H.D., Camacho, D.: A Multi-Objective Graph-based Genetic Algorithm for Image Segmentation. In: Proceedings of the 2014 IEEE International Symposium on INnovations in Intelligent SysTems and Applications (INISTA 2014), pp. 234–241 (2014)
18. Sarath, K.N.V.D., Ravi, V.: Association rule mining using binary particle swarm optimization. Engineering Applications of Artificial Intelligence 26, 1832–1840 (2013)
19. Naveen, N., Ravi, V., Rao, C.R., Sarath, K.N.V.D.: Rule extraction using firefly optimization: Application to banking. In: IEEE International Conference on Industrial Engineering and Engineering Management (2012)
20. Maheshkumar, Y., Ravi, V., Abraham, A.: A particle swarm optimization-threshold accepting hybrid algorithm for unconstrained optimization. Neural Network World, 191–221 (2013)
21. Yang, X.S.: Firefly Algorithm, Stochastic test functions and design optimization. International Journal of Bio-Inspired Computation 2(2), 78–84 (2010)
22. Dueck, G., Scheuer, T.: Threshold Accepting: A general purpose optimization algorithm appearing superior to simulated annealing. Journal of Computational Physics 90(1), 161–175 (1990)
23. Eberhart, R.C., Kennedy, J.: A new optimizer using particle swarm theory. In: Proceedings of the Sixth International Symposium on Micro Machine and Human Science (Nagoya, Japan), pp. 39–43. IEEE Service Center, Piscataway (1995)
24. Wur, S.Y., Leu, Y.: An effective Boolean algorithm for mining association rules in large databases. In: Proceedings of the 6th International Conferenceon Database Systems for Advanced Applications, pp. 179–186 (1998)
25. Kennedy, J., Eberhart, R.: Particle swarm optimization. In: Proceedings of the IEEE International Conference on Neural Networks, vol. 4, pp. 1942–1948 (1995)
26. Kennedy, J., Eberhart, R.C.: A discrete binary version of the particle swarm algorithm. In: Proceedings of the Conference on Systems, Man, and Cybernetics, pp. 4104–4109 (1997)
27. Li, Y., Ning, P., Wang, X.S., Jajodia, S.: Discovering calendar-based temporal association rules. Data and Knowledge Engineering 44(2), 193–218 (2003)
28. Extended Bakery Dataset, https://wiki.csc.calpoly.edu/datasets/wiki/ExtendedBakery
29. Anonymous Web Dataset, http://archive.ics.uci.edu/ml/datasets/Anonymous+Microsoft+Web+Data

Distance-Based Heuristic in Selecting a DC Charging Station for Electric Vehicles*

Junghoon Lee and Gyung-Leen Park

Dept. of Computer Science and Statistics,
Jeju National University, Republic of Korea
{jhlee,glpark}@jejunu.ac.kr

Abstract. This paper proposes a suboptimal tour-and-charging scheduler for electric vehicles which need to select a DC charging station on their single day trips. As a variant of the traveling salesman problem, the tour scheduler finds a visiting order for a given set of destinations and one of any charging stations. To reduce the search space stemmed from a larger number of candidate stations, our distance-based heuristic finds first the nearest destination from each charging station, and calculates the distance between them. Then, m' out of the whole m candidates will be filtered according to the distance. The reduced number of candidates, namely, m', combined with constraint processing on the waiting time, significantly cuts down the execution time for tour schedule generation. The performance measurement result obtained from a prototype implementation reveals that the proposed scheme just brings at most 4.1 % increase in tour length and its accuracy is at least 0.4 with 5 picks, for the given parameter selection.

Keywords: Electric vehicles, tour scheduler, DC charging, TSP variant, distance-based heuristic.

1 Introduction

EVs (Electric Vehicles) are powered by battery-stored electricity [1]. It makes even the transport system be a part of the power network, as their batteries are mainly charged by the energy provided from the grid. EV-based transport will be highly eco-friendly as EVs create no air pollution, while electricity can be produced even from renewable sources such as wind and sunlight [2]. Here, their wide penetration must be preceded by the construction of charging infrastructure, particularly because their driving range is quite short and charging time is much longer than gasoline-powered vehicles. In the early stage charging infrastructure, AC chargers, by which it takes about 6 hours to fully charge an EV battery, are dominating, as they are cheap and pose less burden on the grid

---

* This research was financially supported by the Ministry of Knowledge Economy (MKE), Korea Institute for Advancement of Technology (KIAT) through the Inter-ER Cooperation Projects.

M.N. Murty et al. (Eds.): MIWAI 2014, LNAI 8875, pp. 47–56, 2014.

[3]. However, their long charging time inherently prevents users from readily purchasing EVs.

Gradually, DC chargers are more preferred as they can reduce charging time to 30 minutes [4]. They are quite expensive and may bring a sharp increase in the energy consumption to the grid, especially when many EVs are plugged-in to the grid at the same time [5]. Moreover, fast charging 2 or more times a day may shorten the battery life. However, the convenience brought by the reduced charging time outweighs such problems. In addition, as the daily driving distance is usually covered by overnight AC charging, EVs need to be charged at most once with a DC charger during the daytime. Now, EVs driving beyond the driving range are required to consider where to charge. It is desirable for them to decide the driving route and charging station before they start a trip, possibly making a reservation of a charger. This problem becomes complex when an EV visits multiple destinations like in rent-a-car tours, and the computation time gets longer beyond the tolerable bound.

For the given set of DC chargers, how to select one can be considered a variant of the well-known TSP (Traveling Salesman Problem), which is known to be one of the most time-intensive applications [6]. Basically, it decides the visiting order for n fixed destinations, and many researchers have developed excellent algorithms. However, in our problem, $(n + 1)$-th destination is not given and there are m candidates. Intuitively, it is necessary to add each candidate to the destination set and run the TSP solver one by one to find the best solution. Hence, the time complexity reaches $O(m \times (n + 1)!)$. Here, if we exclude those chargers having little possibility to be selected, we can cut down the response time. Moreover, we can further improve the computation speed by investigating only a set of promising m' (not m) chargers. According to our observation for the problem, a DC charger near a destination is highly likely to reduce the total tour distance and time.

In this regard, this paper designs a distance-based heuristic in selecting a DC charger to make a tour-and-charging schedule for a given set of destinations. A classic computer algorithm is customized to this problem and integrated into our smart grid framework. The information server manages all necessary status records, tracks the current location of each EV, and finally provides the requested operation result to the mobile terminals. Importantly, many articifial intelligence techniques enrich the EV information service [7]. Here, the system design is targeting at a real-life road network and the geographic distribution of chargers and tour spots on Jeju city, Republic of Korea.

2   Related Work

Especially for the fleet management, EV planning is more important for better availability and service ratio. [8] designs an information service architecture for both prior and en-route planning. It focuses on economic itinerary planning based on the retrieval of real-time charging station information. It is built upon location and SoC (State of Charge) tracking for EVs as well as status monitoring for charging stations and road conditions. The global server estimates the battery consumption along the route using previous trip reports and searches the nearest charging station from its local database, when the battery shortage is foreseen. Here, the authors develop a battery consumption model consisting of not only the force factor for an EV to move at a given speed but also the necessary amount of power to overcome the resistance force. Out of several routes recommended by the server, a driver can choose and load the best itinerary to his or her smart phone, which also collects the trip records and reports to the server when returning to the company.

For charging-combined routing, it is necessary to search and reserve a charging station on the way to a destination. [9] presents a routing mechanism in which a broker selects the best charging station for the driver to reserve. The broker considers resource availability, location convenience, price signal change, and others, contacting with various information sources. For the reservation service, each charging station supports the coordination among multiple charging requests over the time slots for both predictable operations and peak-load reduction. In addition, City Trip Planner can create personalized tour routes, providing its service in major cities [10]. For a given set of user-selected tour spots, it maximizes the sum of scores gained by visiting the spot. It automatically inserts tour spots not chosen by users but thought to be preferred. We think that the local search heuristic taken by this approach is worth being considered also for EV tour services.

Not just restricted to charging and tour planning, EVs can bring more profits to EV owners and grids of different levels [11]. Basically, an efficient charging schedule can significantly reduce the fueling cost, considering day-ahead energy prices. The Danish Edison project investigates various EV application areas including a design of EV aggregators operating under the control of mathematical prediction models of driving patterns and charging demand [12]. Those models identify the available charging interval, pursuing cost optimization. In addition, DC charging allows EVs to be charged quickly, namely, within tens of minutes, and even sell electricity back to the grid on peak intervals based on the V2G (Vehicle-to-Grid) technologies [13]. Interestingly, integrating more renewable energies is an attractive advantage of EV charging infrastructure. In [14], the control agent tries to capture as much energy available from renewable sources as possible, maintaining power output as stable as possible. It stores surplus energy to the second-life batteries for better energy efficiency.

Our research team has been developing a tour-and-charging scheduler for EVs which want to visit multiple destinations, considering various options and requirements specified by EV rent-a-car users. The first step of this service design is to trace the SoC change along the tour route [15]. Our team is also refining a battery consumption model for the major roads in Jeju City. The designed scheduler does not only reduce the tour length but also avoids the waiting time, making the tour time overlap EV charging as much as possible. Moreover, a set of system-recommended tour spots facilitating chargers and hosting other activities can eliminate the waiting time. Instead of waiting for batteries to be charged, the tourist can take other actions like dining. In the service implementation, we take both the exhaustive search for accuracy and the genetic algorithm for responsiveness. In addition, one of our previous work has designed a DC charger selection scheme based on the sequential distance from the starting point [16].

3   DC Charging Station Selection

In the target city, namely, Jeju City, Republic of Korea, 21 DC chargers are currently installed. DC chargers evenly distribute over the whole island area and the number of DC chargers keeps increasing. To decide a TSP path, it is necessary to know the cost of every destination pair. Our service allows tourists to select the tour spots only from this fixed set for manageable computation time. This restriction makes it possible to precalculate all inter-destination distances by conventional point-to-point shortest path algorithms such as Dijkstra or A*, regardless of their execution time. The north region has the largest population and many facilities such as the international airport and hotels, making many tours start from here. The in-town chargers are mainly used by local residents during day and night time.

In Jeju City, having the perimeter of 250 km and a variety of tourist attractions, the statistics tell that the daily tour length usually falls into the range of 100 to 150 km. Hence, one charging is essential during the trip, but it may extend the tour length and time. Straightforwardly, given a set of user-selected n destinations, every DC charger is added to the tour set and the one making the tour length smallest will survive the competition. With the exhaustive search, the tour scheduler investigates all feasible schedules. Here, the first level of the search tree has m subtrees, each of which has $n!$ leaves, making the depth of the tree $(n+1)$. Lke other optimization technologies, reaching a leaf node, a complete schedule is built and its cost is evaluated according to the given criteria. If its tour length is less than the current best and it does not violate the constraint, this schedule will replace the current best.

If a schedule makes passengers to wait somewhere on the route to charge the battery with slow chargers, they cannot accept it. To remove such a schedule, the search procedure estimates the SoC change along the route [17]. The Soc decreases according to the distance the EV has taken. If SoC drops below 0 somewhere on the sequence specified in the schedule, the schedule will be automatically discarded without further evaluation. In the process of the search space traversal during which each node is expanded one by one, a subsequence from the root, namely, the subpath from the starting point to the currently expanding node, can be already larger than the current best. Then, it is necessary to stop the expansion immediately by pruning the branch. In the tour schedule, the number of selected tour spots, namely, n, is usually less than 10 in most tour cities, and the tour schedule can be generated within a tolerable bound even with an average-performance PC. In spite of much achievable improvement in response time, a large m can make the execution time too much.

Our main idea lies in achieving an acceptable response time even if m gets larger according to the installation of more DC chargers. The responsiveness comes with a small accuracy loss as the scheduling procedure investigates just m' out of m chargers. m' is a tunable parameter and the number of chargers to investigate. The problem definition and the main idea are both illustrated in Figure 1, where D_1, D_2, D_3, and D_4 are destinations an EV wants to visit and tour length will be decided by the visiting order. We assume that the EV starts from and returns to D_1. There are 5 DC charging stations from C_1 to C_5 and the EV needs to be charged once during the tour and thus it must additionally visit one of them. In addition, the pure tour length decided by a TSP solver for $\{D_1, D_2, D_3, D_4\}$ is defined as TSP length. It must be distinguished from the travel distance which includes the addition of some C_i.

A preprocessing procedure has calculated the battery consumption and the distance for each pair of two destinations and also for each pair of a destination and a charging station. By a legacy TSP solver, we can traverse the search space and find an optimal schedule for n destinations based on the inter-destination cost matrix. Hence, the problem is reduced to finding C_k which minimizes the tour length of a schedule for $\{D_1, D_2, .., D_n, C_k\}$ without making its waiting time nonzero. Basically, the TSP sovler can be invoked m times, namely, for $\{D_1, D_2, .., D_n, C_1\}$, $\{D_1, D_2, .., D_n, C_2\}$, ..., $\{D_1, D_2, .., D_n, C_m\}$ to find the optimal schedule. However, if m is large, the execution time is extended too much, as the TSP solver basically belongs to the $O(n!)$ problem category. If we can find a reasonable quality solution with m' candidates ($m' \ll m$), the search space will be significantly cut down, and $\frac{m'}{m}$ is the speedup ratio.

Our heuristic of selecting m' candidates can be better explained with the example of Figure 1. An addition of a charging station to the tour inevitably increases the tour length and time. However, if a station on the tour route is selected, the tour distance is hardly affected. We cannot know which link (pair of two destinations) will be included in the final schedule until the time-intensive TSP solver completes the schedule. On the contrary, if a station near any destination is selected, it is highly likely not to increase the tour length. Hence, in Figure 1, for each charging station, it is necessary to find the closest

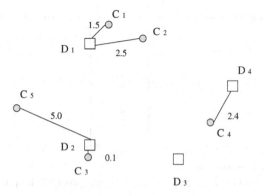

Fig. 1. Distance-based heuristic

destination. Here, C_1 and C_2 are commonly closest to D_1. As the cost matrix from a station to every destination is also given, this step just finds the D_i having the smallest distance. Now, excluding the start point, which is the first and last destination of a sequence and thus charging near this point may make the waiting time nonzero, we can find the charging station having the smallest distance. In this figure, C_3 has 0.1 and will be chosen to the m' candidates first. If m' is 2, C_4 will be selected.

4   Performance Measurement

This section first implements a prototype of the proposed tour scheduler and evaluates its performance. An optimal scheduler, which investigates all feasible schedules, namely, $m=m'$, is also developed for performance comparison. For simplicity, we assume that there is no queuing delay to focus on the pure effect of DC charger selection, as it can be easily integrated with a charging schedule. When multiple charging requests concentrate on the same time slot, the queuing time must be considered. In selecting m' candidates, those stations already reserved on the estimated arrival time of the EV should be avoided. Main performance metrics include tour length and accuracy. The accuracy means the probability that a scheduler finds the optimal sequence. For 40 destinations, the distance for each pair of them is calculated in priori. $(n-1)$ destinations are randomly picked in addition to the airport, the start point. If a schedule is found for a given set of destinations, it will be regarded as a feasible one. For each parameter setting, 10 sets are generated and the results are averaged.

The first experiment measures the travel distance according to the number of destinations and also the number of picks, while the results are plotted in Figure 2. As shown in Figure 2(a), the experiment changes the number of destinations from 5 to 8. Here, the destination sets having the TSP length of 100.0 to 120.0 km are chosen. For 9 or more destinations, it is hard to find a destination set whose TSP length falls in this range. Figure 2(a) shows 3 curves. As expected,

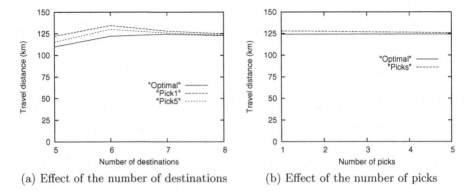

(a) Effect of the number of destinations   (b) Effect of the number of picks

Fig. 2. Effect of the number of picks

the optimal scheme has the smallest travel distance on the whole range. The Pick$n$ curve corresponds to the case m' is n. The performance gap is largest on 5 destinations and gets smaller according to the increase in the number of destinations. With 8 destinations, Pick5 and Pick1 are just 0.7 % and 1.6 % longer than the optimal result, respectively. This behavior can be explained by the observation that with more destinations, the scheduler is more likely to find candidates closer to destinations.

Next, Figure 2(b) shows a deeper investigation result on the effect of the number of picks. Here, the number of destinations is fixed to 7 and the TSP length is made to range from 100.0 to 120.0 km as in the previous experiment. The travel distance of the optimal schedule outperforms others and with more candidates we can expect a better travel distance. In this parameter setting, even with just 1 pick, which corresponds to $\frac{1}{21}$ of the legacy computation time, the travel distance is just 1.6 % longer than the optimal schedule. Furthermore, the gap linearly decreases each time a candidate is added. This result indicates that Pick5 almost always finds the optimal schedule for 10 sets just with the $\frac{5}{21}$ of the execution time.

The second experiment measures the effect of the TSP length to the travel distance. It must be mentioned again that the difference between them comes from the addition of a charging station to the entire tour schedule. Figure 3 shows the travel distance according to the TSP length. Here, for each 10 km interval in the TSP length, for example, from 100.0 to 110.0 km and from 130.0 to 140.0 km, 10 destination sets are selected. The number of destinations is fixed to 7. In Figure 3, the Pick1 graph goes quite higher than the others, especially when the TSP length is over 120.0 km. It reaches 19.3 % for the range of 130.0 to 140.0 km. In selecting a DC charger, Pick1 simply considers the distance from any one of the destinations. If a DC charger close to the start (or last) destination in the sequence is selected, it can lead to the nonzero waiting time. On the contrary, when m' is 5, this effect can be almost completely masked out. As shown in the figure, Pick5 generates a schedule at most 4.1 % longer than the optimal scheme.

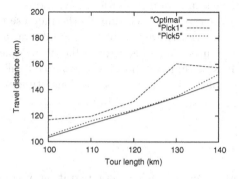

Fig. 3. Effect of the TSP length

The final experiment measures the accuracy according to the number of picks and also the number of destinations, the results being shown in Figure 4. Figure 4(a) plots the measured accuracy according to the number of picks from 1 to 5. In all experiments, the TSP length is made to range from 100.0 to 120.0 km. Each curve corresponds to the respective number of destinations. For all curves, the increase in the number of picks essentially improves the accuracy. With 5 picks, the accuracy is at least 0.5, reaching 0.8 for the cases of 5 and 7 destinations. Even in the case the scheduler fails to find an optimal schedule, its quality is comparable to it. This result indicates that our scheme can find a near optimal schedule with much less and even tunable response time compared with the optimal schedule which may take too much time when there are many charging stations.

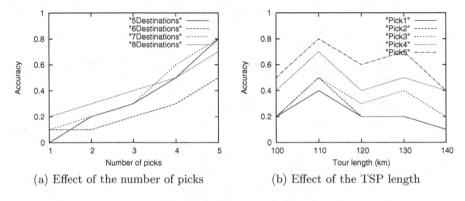

(a) Effect of the number of picks   (b) Effect of the TSP length

Fig. 4. Accuracy analysis

In addition, Figure 4(b) plots the accuracy according to the TSP length. Here, the number of destinations is fixed to 7. This figure includes 5 curves, each of which is associated with each number of picks. Unlike Figure 4(a), Figure 4(b) doesn't seem to be linearly dependent on the TSP length. The destination set specific features, such as tour spot distribution, have more effect on the accuracy. The accuracy tends to deteriorate according to the increase in the TSP length, but this effect is not so vivid. However, during the interval from 110.0 to 130.0 km, we can find the accuracy remains at least 0.6 with 5 picks and this result confirms again that it is possible to obtain a reasonable quality schedule for the whole range of TSP length, with much less response time, compared with the optimal scheme.

5   Conclusions

Modern grids are getting smarter with the integration of computational intelligence supported by information technologies. EVs, still facing an obstacle of long charging time and short driving range towards their wide penetration, can also benefit from intelligent tour planning. For EV rent-a-cars which visit multiple destinations and need to be charged once a day, an efficient selection of a charging station along with a tour schedule can reduce the tour length and save energy. However, we must cope with the time complexity according to the increase in the number of available chargers following the expansion of charging infrastructure. This paper achieves this goal by constraint processing and the reduction of the number of candidate stations from m to m', where m is the number of all stations in the area and m' is that in the selected subset. The subset is built by a heuristic which selects m' candidates based on the distance from each station to any of given destinations.

The performance measurement result, conducted on the real-life environment of the target city, shows that with 5 candidates, the tour length is prolonged just by 0.7 % and finds the optimal schedule with the probability of 0.8, with the speedup of $\frac{5}{21}$. Next, for the given TSP length range, the proposed scheme just brings at most 4.1 % of distance overhead, compared with the optimal scheme. As future work, we are planning to combine with the spatial query processing system to cope with the increase in the number of DC chargers. By filtering out those chargers far away from selected destinations, we can further reduce the size of m.

References

1. Timpner, J., Wolf, L.: Design and evaluation of charging station scheduling strategies for electric vehicles. IEEE Transactions on Intelligent Transportation Systems 15(2), 579–588 (2014)
2. Mischinger, S., Hennings, W., Strunz, K.: Integration of surplus wind energy by controlled charging of electric vehicles. In: 3rd IEEE PES Innovative Smart Grid Technologies Europe (2012)
3. Botsford, C., Szczepanek, A.: Fast charging vs. slow charging: Pros and cons for the new age of electric vehicles. In: International Battery Hybrid Fuel Cell Electric Vehicle Symposium (2009)
4. Veneri, O., Capasso, C., Ferraro, L., Pizzo, A.: Performance analysis on a power architecture for EV ultra-fast charging stations. In: International Conference on Clean Electrical Power, pp. 183–188 (2013)
5. Vedova, M., Palma, E., Facchinetti, T.: Electric load as real-time tasks: An application of real-time physical systems. In: International Wireless Communications and Mobile Computing Conference, pp. 1117–1123 (2011)
6. Shim, V., Tan, K., Tan, K.: A hybrid estimation of distribution algorithm for solving the multi-objective multiple traveling salesman problem. In: IEEE World Congress on Computational Intelligence (2012)
7. Ramchrun, S., Vytelingum, R., Rogers, A., Jennings, N.: Putting the 'smarts' into the smart grid: A grand challenge for artificial intelligence. Communications of the ACM 55(4), 86–97 (2012)
8. Mehar, S., Remy, G.: EV-planning: Electric vehicle itinerary planning. In: International Conference on Smart Communications in Network Technologies (2013)
9. Bessler, S., Grønbæk, J.: Routing EV users towards an optimal charging plan. In: International Battery, Hybrid and Fuel Cell Electric Vehicle Symposium (2012)

10. Vansteenwegen, P., Souffriau, W., Berghe, G., Oudheusden, D.: The City trip planner: An expert system for tourists. Expert Systems with Applications 38, 6540–6546 (2011)
11. Andrandt, A., Andersen, P., Pedersen, A., You, A., Poulsen, B., O'Cornel, N., Østergaard, J.: Prediction and optimization methods for electric vehicle charging schedules in the Edison project. IEEE Transactions on Smart Grid, 111–119 (2011)
12. Ortega-Vazquez, M., Bouffard, F., Silva, V.: Electric vehicle aggregator/system operator coordination for charging scheduling and services procurement. IEEE Transactions on Power Systems 28(2), 1806–1815 (2013)
13. Kisacikoglu, M., Ozpineci, B., Tolbert, L.: EV/PHEV bidirectional charger assessment for V2G reactive power operation. IEEE Transactions on Power Electronics 28(12), 5717–5727 (2013)
14. Hamidi, A., Weber, L., Nasiri, A.: EV charging station integrating renewable energy and second-life battery. In: International Conference on Renewable Energy Research and Applications, pp. 1217–1221 (2013)
15. Lee, J., Park, G.: A tour recommendation service for electric vehicles based on a hybrid orienteering model. In: ACM Symposium on Applied Computing, pp. 1652–1654 (2013)
16. Lee, J., Park, G.: DC charger selection scheme for electric vehicle-based tours visiting multiple destinations. In: ACM Research in Applied Computation Symposium (to appear, 2014)
17. Lee, J., Park, G.-L.: Design of a multi-day tour-and-charging scheduler for electric vehicles. In: Ramanna, S., Lingras, P., Sombattheera, C., Krishna, A. (eds.) MIWAI 2013. LNCS, vol. 8271, pp. 108–118. Springer, Heidelberg (2013)

Automated Reasoning in Deontic Logic⋆

Ulrich Furbach[1], Claudia Schon[1] and Frieder Stolzenburg[2]

[1] Universität Koblenz-Landau, Koblenz, Germany
{uli,schon}@uni-koblenz.de
[2] Harz University of Applied Sciences, Wernigerode, Germany
fstolzenburg@hs-harz.de

Abstract. Deontic logic is a very well researched branch of mathematical logic and philosophy. Various kinds of deontic logics are discussed for different application domains like argumentation theory, legal reasoning, and acts in multi-agent systems. In this paper, we show how standard deontic logic can be stepwise transformed into description logic and DL-clauses, such that it can be processed by Hyper, a high performance theorem prover which uses a hypertableau calculus. Two use cases, one from multi-agent research and one from the development of normative system are investigated.

Keywords: Deontic Logic, Automated Theorem Proving, Description Logics.

1 Introduction

Deontic logic is a very well researched branch of mathematical logic and philosophy. Various kinds of deontic logics are discussed for different application domains like argumentation theory, legal reasoning, and acts in multi-agent systems [11]. Recently there also is growing interest in modelling human reasoning and testing the models with psychological findings. Deontic logic is an obvious tool to this end, because norms and licenses in human societies can be described easily with it. For example in [9] there is a discussion of some of these problems including solutions with the help of deontic logic. There, the focus is on using deontic logic for modelling certain effects, which occur in human reasoning, e.g. the Wason selection task or Byrne's suppression task.

This paper concentrates on automated reasoning in standard deontic logic (SDL). Instead of implementing a reasoning system for this logic directly, we rather rely on existing methods and systems. Taking into account that SDL is just the modal logic K with a seriality axiom, we show that deontic logic can be translated into description logic $\mathcal{ALC}$. The latter can be transformed into so called DL-clauses, which is a special normal form with clauses consisting of implications where the body is, as usual, a conjunction of atoms and the head is a disjunction of literals. These literals can be atoms or existential quantified expressions.

⋆ Work supported by DFG grants FU 263/15-1 and STO 421/5-1 'Ratiolog'.

M.N. Murty et al. (Eds.): MIWAI 2014, LNAI 8875, pp. 57–68, 2014.

DL-clauses can be decided by the first-order reasoning system Hyper [22], which uses the hypertableau calculus from [4]. In Sections 2 and 3 we shortly depict this workflow, and in Section 4 we demonstrate the use of our technique with the help of two problems from the literature, one from multi-agent research and the other one from testing normative systems. We choose these examples, because they hopefully document the applicability of reasoning of SDL in various areas of AI research.

2   Deontic Logic as Modal Logic **KD**

We consider a simple modal logic which consists of propositional logic and the additional modal operators $\Box$ and $\Diamond$. Semantics are given as possible world semantics, where the modal operators $\Box$ and $\Diamond$ are interpreted as quantifiers over possible worlds. Such a possible world is an assignment, which assigns truth values to the propositional variables. An interpretation connects different possible worlds by a (binary) reachability relation R. The $\Box$-operator states that a formula has to hold in all reachable worlds. Hence if v and w are worlds, we have

$$w \models \Box P \qquad \text{iff} \qquad \forall v : R(w, v) \to v \models P$$

Standard deontic logic (SDL) is obtained from the well-known modal logic K by adding the seriality axiom D:

$$\mathsf{D} : \quad \Box P \to \Diamond P$$

In this logic, the $\Box$-operator is interpreted as 'it is obligatory that' and the $\Diamond$ as 'it is permitted that'. The $\Diamond$-operator can be defined by the following equivalence:

$$\Diamond P \equiv \neg \Box \neg P$$

The additional axiom D: $\Box P \to \Diamond P$ in SDL states that, if a formula has to hold in all reachable worlds, then there exists such a world. With the deontic reading of $\Box$ and $\Diamond$ this means: Whenever the formula P ought to be, then there exists a world where it holds. In consequence, there is always a world, which is ideal in the sense, that all the norms formulated by 'the ought to be'-operator hold.

SDL can be used in a natural way to describe knowledge about norms or licenses. The use of conditionals for expressing rules which should be considered as norms seems likely, but holds some subtle difficulties. If we want to express that *if P then Q* is a norm, an obvious solution would be to use

$$\Box(P \to Q)$$

which reads *it is obligatory that Q holds if P holds*. An alternative would be

$$P \to \Box Q$$

meaning *if P holds, it is obligatory that Q holds*. In [21] there is a careful discussion which of these two possibilities should be used for conditional norms. The first one has severe disadvantages. The most obvious disadvantage is, that P together with $\Box(P \to Q)$ does not imply $\Box Q$. This is why we prefer the latter method, where the $\Box$-operator is in the conclusion of the conditional. We will come back to this point in Subsection 4.1 where we consider several formalization variants of the well-known problem of contrary-to-duty-obligations. For a more detailed discussion of such aspects we refer to [10].

3   Automated Reasoning for Deontic Logic

Deontic logic is the logic of choice when formalizing knowledge about norms like the representation of legal knowledge. However, there are only few automated theorem provers specially dedicated for deontic logic and used by deontic logicians (see [1,3]). Nonetheless, numerous approaches to translate modal logics into (decidable fragments of) first-order predicate logics are stated in the literature. A nice overview including many relevant references is given in [19].

In this paper, we describe how to use the Hyper theorem prover [22] to handle deontic logic knowledge bases. These knowledge bases can be translated efficiently into description logic formulae. Hyper is a theorem prover for first-order logic with equality. In addition to that, Hyper is a decision procedure for the description logic $\mathcal{SHIQ}$ [6].

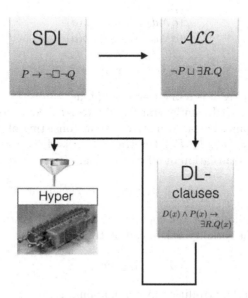

Fig. 1. From SDL to Hyper. Note that concept D occurring in the DL-clauses is an auxiliary concept.

In Figure 1, we depict the entire workflow from a given SDL knowledge base to the final input into the Hyper theorem prover. In the following, we describe these three steps in more detail.

3.1   Transformation from Deontic Logic into $\mathcal{ALC}$

First, we will show how to translate SDL knowledge bases into $\mathcal{ALC}$ knowledge bases. An $\mathcal{ALC}$ knowledge base consists of a TBox and an ABox. The TBox (terminological box) gives information about concepts occurring in the domain of interest and describes concept hierarchies. The ABox (assertional box) introduces individuals and states, to which concepts the individuals belong and how they are interconnected via relations called roles . The ABox contains assertional knowledge and can be seen as the representation of a state of the world. We do not give the syntax and semantics of $\mathcal{ALC}$ here and refer the reader to [2].

There is a strong connection between modal logic and the description logic $\mathcal{ALC}$. As shown in [18], the description logic $\mathcal{ALC}$ is a notational variant of the modal logic K_n. Therefore any formula given in the modal logic K_n can be translated into an $\mathcal{ALC}$ concept and vice versa. Since we are only considering a modal logic as opposed to a multimodal logic, we will omit the part of the translation handling the multimodal part of the logic. Mapping φ translating from modal logic K formulae to $\mathcal{ALC}$ concepts is inductively defined as follows:

$$\varphi(\top) = \top$$
$$\varphi(\bot) = \bot$$
$$\varphi(b) = b$$
$$\varphi(\neg c) = \neg\varphi(c)$$
$$\varphi(c \wedge d) = \varphi(c) \sqcap \varphi(d)$$
$$\varphi(c \vee d) = \varphi(c) \sqcup \varphi(d)$$
$$\varphi(\Box c) = \forall r.\varphi(c)$$
$$\varphi(\Diamond c) = \exists r.\varphi(c)$$

Note that the mapping φ is a one-to-one mapping.

Formulae given in SDL can be translated into $\mathcal{ALC}$ concepts using the above introduced φ mapping. For a normative system consisting of the set of deontic logic formulae $\mathcal{N} = \{F_1, \ldots, F_n\}$ the translation is defined as the conjunctive combination of the translation of all deontic logic formulae in $\mathcal{N}$:

$$\varphi(\mathcal{N}) = \varphi(F_1) \sqcap \ldots \sqcap \varphi(F_n)$$

Note that $\varphi(\mathcal{N})$ does not yet contain the translation of the seriality axiom. As shown in [12] the seriality axiom can be translated into the following TBox:

$$\mathcal{T} = \{\top \sqsubseteq \exists r.\top\}$$

with r the atomic role introduced by the mapping φ.

For our application, the result of the translation of a normative system $\mathcal{N}$ and the seriality axiom is an $\mathcal{ALC}$ knowledge base $\Phi(\mathcal{N}) = (\mathcal{T}, \mathcal{A})$, where the TBox $\mathcal{T}$

consists of the translation of the seriality axiom and the ABox $\mathcal{A} = \{(\varphi(\mathcal{N}))(a)\}$ for a new individual a. In description logics performing a satisfiability test of a concept C w.r.t. a TBox is usually done by adding a new individual a together with the ABox assertion $C(a)$. For the sake of simplicity, we do this construction already during the transformation of Φ by adding $(\varphi(\mathcal{N}))(a)$ to the ABox.

An advantage of the translation of deontic logic formulae into an $\mathcal{ALC}$ knowledge base is the existence of a TBox in $\mathcal{ALC}$. This makes it possible to add further axioms to the TBox. For example we can add certain norms that we want to be satisfied in all reachable worlds into the TBox.

3.2 Translation from $\mathcal{ALC}$ into DL-Clauses

Next we transform the $\mathcal{ALC}$ knowledge base into so called DL-clauses introduced in [15] which represent the input format for the Hyper theorem prover.

DL-clauses are constructed from so called *atoms*. An atom is of the form $b(s)$, $r(s,t)$, $\exists r.b(s)$ or $\exists r.\neg b(s)$ for b an atomic concept and s and t individuals or variables. They are universally quantified implications of the form

$$\bigwedge_{i=1}^{m} u_i \rightarrow \bigvee_{j=1}^{n} v_j$$

where the u_i are atoms of the form $b(s)$ or $r(s,t)$ and the v_j may be arbitrary DL-clause atoms, i.e. including existential quantification, with $m, n \geq 0$.

Comparing the syntax of DL-clauses to the syntax of first order logic clauses written as implications, the first obvious difference is the absence of function symbols. The second difference is the fact, that in DL-clauses all atoms are constructed from unary or binary predicates. The most interesting difference however is the fact, that the head of a DL-clause is allowed to contain atoms of the form $\exists r.b(s)$.

The basic idea of the translation of an $\mathcal{ALC}$ knowledge base into DL-clauses is that the subsumption in a TBox assertion is interpreted as an implication from the left to the right side. Further concepts are translated to unary and roles to binary predicates. Depending on the structure of the assertion, auxiliary concepts are introduced. For example the TBox axiom

$$d \sqsubseteq \exists r.b \sqcup \forall r.c$$

corresponds to the following DL-clause

$$d(x) \wedge r(x,y) \rightarrow c(y) \vee \exists r.b(x)$$

For detailed definitions of both syntax and semantics of DL-clauses and the translation into DL-clauses, we refer the reader to [15]. The translation preserves equivalence, avoids an exponential blowup by using a well-known structural transformation [17] and can be computed in polynomial time. In the following, for an $\mathcal{ALC}$ knowledge base $\mathcal{K} = (\mathcal{T}, \mathcal{A})$, the corresponding set of DL-clauses is denoted by $\omega(\mathcal{K})$.

3.3   Reasoning Tasks

With the help of Hyper, we can solve several interesting reasoning tasks:

- **Consistency checking of normative systems:** In practice, normative systems can be very large. Therefore it is not easy to see, if a given normative system is consistent. The Hyper theorem prover can be used to check consistency of a normative system $\mathcal{N}$. We first translate $\mathcal{N}$ into an $\mathcal{ALC}$ knowledge base $\Phi(\mathcal{N})$, then translate $\Phi(\mathcal{N})$ into the set $\omega(\Phi(\mathcal{N}))$ of DL-clauses. Then we can check the consistency of $\omega(\Phi(\mathcal{N}))$ using Hyper.
- **Evaluation of normative systems:** Given several normative systems, we use Hyper to find out for which normative system guarantees a desired outcome is guaranteed.
- **Independence checking:** Given a normative system $\mathcal{N}$ and a formula F representing a norm, we can check whether F is independent from $\mathcal{N}$. If F is independent from $\mathcal{N}$, then F is not a logical consequence of $\mathcal{N}$.

In Section 4, we will give detailed examples for those tasks. Subsection 4.1 gives an example for a consistency check of a normative system and illustrates how the independence of a formula from a normative system can be decided. In Subsection 4.2, we use an example from multi-agent systems to show how to evaluate normative systems.

4   Applications

The literature on deontic logic deals with numerous small but nonetheless interesting examples. They are mostly used to show typical problems or special features of the logic under consideration (cf. [10]). In Subsection 4.1, we deal with one of these examples. In Subsection 4.2, we formalize a 'real-life' problem from multi-agent research.

4.1   Contrary-to-duty Obligations

Let us now consider consistency testing of normative systems and independence checking. As an example, we examine the well-known problem of *contrary-to-duty obligations* introduced in [8]:

(1) a ought not steal.
(2) a steals.
(3) If a steals, he ought to be punished for stealing.
(4) If a does not steal, he ought not be punished for stealing.

Table 1 shows three different formalizations of this problem. Those formalizations are well-known from the literature [5,14,13,21]:

Table 1. Formalizations of the *contrary-to-duty obligation* introduced in [8]

	$\mathcal{N}_1$	$\mathcal{N}_2$	$\mathcal{N}_3$
(1)	$\Box\neg s$	$\Box\neg s$	$\Box\neg s$
(2)	s	s	s
(3)	$s \to \Box p$	$\Box(s \to p)$	$s \to \Box p$
(4)	$\Box(\neg s \to \neg p)$	$\Box(\neg s \to \neg p)$	$\neg s \to \Box\neg p$

Table 2. Translation of the normative system $\mathcal{N}_1$ into $\varphi(\mathcal{N}_1)$

$\mathcal{N}_1$ (in Deontic Logic)	$\varphi(\mathcal{N}_1)$
$\Box\neg s$	$\forall r.\neg s$
s	s
$s \to \Box p$	$\neg s \sqcup \forall r.p$
$\Box(\neg s \to \neg p)$	$\forall r.(s \sqcup \neg p)$

Consistency Testing of Normative Systems. The contrary-to-duty obligation formalized above is a very small example. In practice, normative systems can be rather complex. This makes it difficult to see if a normative system is consistent. We will show how to use the Hyper theorem prover to check the consistency of a given normative system.

As an example, we consider formalization $\mathcal{N}_1$ given in Table 1 which, according to [21], is inconsistent. We will use Hyper to show this inconsistency. For this, we first translate normative system $\mathcal{N}_1$ into an $\mathcal{ALC}$ knowledge base $\Phi(\mathcal{N}_1)$. Table 2 shows $\varphi(\mathcal{N}_1)$.

To perform the satisfiability test, we transform the description logic representation $\Phi(\mathcal{N}_1)$ into a set of DL-clauses $\omega(\Phi(\mathcal{N}_1))$. Hyper constructs a hypertableau for $\omega(\Phi(\mathcal{N}_1))$. This hypertableau is closed and therefore we can conclude that $\mathcal{N}_1$ is inconsistent.

Independence Checking. Normative System $\mathcal{N}_2$ given in Table 1 is consistent. However it has another drawback: The different formulae in this formalization are not independent from another. Formula (3) is a logical consequence of (1), because $\Box(s \to p) \equiv \Box(\neg s \vee p)$ (definition of $\to$) which clearly is implied by the (subsuming) formula (1) $\Box\neg s$. We can use Hyper to show this by transforming the problem into a satisfiability test. For this, we remove formula (3) from $\mathcal{N}_2$ and add its negation $\neg\Box(s \to p)$ to $\mathcal{N}_2$. If the resulting normative system is inconsistent, we can conclude, that formula (3) is not independent from the other formulae in $\mathcal{N}_2$.

The problem of independence of formulae given in a normative system is interesting in practice as well. If an existing normative system is extended with some new formulae, it is interesting to know, whether the new formulae are independent from the original normative system. This can be checked automatically using Hyper as described above.

In the same way, we can show, that formula (4) is not independent from $\mathcal{N}_3$. Note that only this normative system is both consistent and represents all conditionals carefully, i.e. with formulae of the form $P \to \Box Q$ (cf. Section 2). Only for this formalization we have: If a steals in the actual world, a will be punished in the corresponding reachable ideal world.

4.2   An Example from Multi-agent Systems

In multi-agent systems, there is a relatively new area of research, namely the formalization of 'robot ethics'. It aims at defining formal rules for the behavior of agents and to prove certain properties. As an example consider Asimov's laws, which aim at regulating the relation between robots and humans. In [7] the authors depict a small example of two surgery robots obeying ethical codes concerning their work. These codes are expressed by means of MADL, which is an extension of standard deontic logic with two operators. In [16] an axiomatization of MADL is given. Further it is asserted, that MADL is not essentially different from standard deontic logic. This is why we use SDL to model the example.

Formalization in SDL. In our example, there are two robots $ag1$ and $ag2$ in a hospital. For sake of simplicity, each robot can perform one specific action: $ag1$ can terminate a person's life support and $ag2$ can delay the delivery of pain medication. In [7] four different ethical codes J, $J^\star$, O and $O^\star$ are considered:

- "If ethical code J holds, then robot $ag1$ ought to take care, that life support is terminated." This is formalized as:

$$J \to \Box act(ag1, term)$$

- "If ethical code $J^\star$ holds, then code J holds, and robot $ag2$ ought to take care, that the delivery of pain medication is delayed." This is formalized as:

$$J^\star \to J \wedge J^\star \to \Box act(ag2, delay)$$

- "If ethical code O holds, then robot $ag2$ ought to take care, that delivery of pain medication is not delayed." This is formalized as:

$$O \to \Box \neg act(ag2, delay)$$

- "If ethical code $O^\star$ holds, then code O holds, and robot $ag1$ ought to take care, that life support is not terminated." This is formalized as:

$$O^\star \to O \wedge O^\star \to \Box \neg act(ag1, term)$$

Further we give a slightly modified version of the evaluation of the robot's acts given in [7], where $(+!!)$ describes the most and $(-!!)$ the least desired outcome. Note that terms like $(+!!)$ are just propositional atomic formulae here.

$$act(ag1, term) \wedge \quad act(ag2, delay) \to (-!!) \tag{1}$$

$$act(ag1, term) \wedge \neg act(ag2, delay) \to (-!) \tag{2}$$

$$\neg act(ag1, term) \wedge \quad act(ag2, delay) \to (-) \tag{3}$$

$$\neg act(ag1, term) \wedge \neg act(ag2, delay) \to (+!!) \tag{4}$$

Table 3. Translation of the normative system $\mathcal{N}$ into $\varphi(\mathcal{N})$

Deontic Logic	$\mathcal{ALC}$
$J \to \Box act(ag1, term)$	$\neg J \sqcup \forall r.act(ag1, term)$
$J^* \to J \wedge J^* \to \Box act(ag2, delay)$	$(\neg J^* \sqcup J) \sqcap (\neg J^* \sqcup \forall r.act(ag2, delay))$
$O \to \Box \neg act(ag2, delay)$	$\neg O \sqcup \forall r.\neg act(ag2, delay)$
$O^* \to O \wedge O^* \to \Box \neg act(ag1, term)$	$(\neg O^* \sqcup O) \sqcap (\neg O^* \sqcup \forall r.\neg act(ag1, term))$
$act(ag1, term) \wedge \ \ act(ag2, delay) \to (-!!)$	$\neg(act(ag1, term) \sqcap \ \ act(ag2, delay)) \sqcup (-!!)$
$act(ag1, term) \wedge \neg act(ag2, delay) \to (-!)$	$\neg(act(ag1, term) \sqcap \neg act(ag2, delay)) \sqcup (-!)$
$\neg act(ag1, term) \wedge \ \ act(ag2, delay) \to (-)$	$\neg(\neg act(ag1, term) \sqcap \ \ act(ag2, delay)) \sqcup (-)$
$\neg act(ag1, term) \wedge \neg act(ag2, delay) \to (+!!)$	$\neg(\neg act(ag1, term) \sqcap \neg act(ag2, delay)) \sqcup (+!!)$

These formulae evaluate the outcome of the robots' actions. It makes sense to assume, that this evaluation is effective in all reachable worlds. This is why we add formulae stating that formulae (1)–(4) hold in all reachable worlds. For example, for (1) we add:

$$\Box(act(ag1, term) \wedge act(ag2, delay) \to (-!!)) \tag{5}$$

Since our example does not include nested modal operators, the formulae of the form (5) are sufficient to spread the evaluation formulae to all reachable worlds. The normative system $\mathcal{N}$ formalizing this example consists of the formalization of the four ethical codes and the formulae for the evaluation of the robots actions.

Reduction to a Satisfiability Test A possible query would be to ask, if the most desirable outcome $(+!!)$ will come to pass, if ethical code O^* is operative. This query can be translated into a satisfiability test: If

$$\mathcal{N} \wedge O^* \wedge \Diamond \neg(+!!)$$

is unsatisfiable, then ethical code O^* ensures outcome $(+!!)$.

Translation into Description Logic. As described in Section 3.1, we translate normative system $\mathcal{N}$ given in the previous section into an $\mathcal{ALC}$ knowledge base $\Phi(\mathcal{N}) = (\mathcal{T}, \mathcal{A})$. Table 3 shows the result of translating $\mathcal{N}$ into $\varphi(\mathcal{N})$.

We further add the following two assertions to the ABox $\mathcal{A}$:

$$O^*(a)$$
$$\exists r.\neg(+!!)(a)$$

Next we translate the knowledge base into DL-clauses and use Hyper to test the satisfiability of the resulting set of DL-clauses. Using further satisfiability tests, we can show, that ethical codes J, J^* or O are not sufficient to guarantee the most desired outcome $(+!!)$.

Formalization in Description Logic Using a TBox. In the formalization given in the previous subsection, we added formulae stating that the evaluation of the agents' actions holds in all worlds, which are reachable in one step, see (5) for an example. In our case it is sufficient to add formulae of the form (5) because the formalization does not include nested modal operators. In general it is desirable to express that those formulae hold in *all* reachable worlds including worlds reachable in more than one step. However this would mean to either add infinitely many formulae or to use a universal modality, i.e. the reflexive-transitive closure of the respective simple modality.

In description logics we can use a more elegant way to formalize that all worlds are supposed to fulfill certain formulae. Description logic knowledge bases contain a TBox including the terminological knowledge. Every individual is supposed to fulfill the assertions given in the TBox. Hence, we can add the formulae stating the evaluation of the agents' actions into the TBox. For this, we reformulate implication ($\rightarrow$) by subsumption ($\sqsubseteq$). We model the deontic logic formulae given in Table 3 by the following TBox $\mathcal{T}$:

$$\top \sqsubseteq \exists r.\top$$
$$J \sqsubseteq \forall r.act(ag1, term)$$
$$J^\star \sqsubseteq J$$
$$J^\star \sqsubseteq \forall r.act(ag2, delay)$$
$$O \sqsubseteq \forall r.\neg act(ag2, delay)$$
$$O^\star \sqsubseteq O$$
$$O^\star \sqsubseteq \forall r.\neg act(ag1, term)$$
$$act(ag1, term) \sqcap act(ag2, delay) \sqsubseteq (-!!)$$
$$act(ag1, term) \sqcap \neg act(ag2, delay) \sqsubseteq (-!)$$
$$\neg act(ag1, term) \sqcap act(ag2, delay) \sqsubseteq (-)$$
$$\neg act(ag1, term) \sqcap \neg act(ag2, delay) \sqsubseteq (+!!)$$

Reduction to a Satisfiability Test Like in the previous section, we now want to know, if the most desirable outcome $(+!!)$ will come to pass, if ethical code $O^\star$ is operative. We perform this test by checking the satisfiability of the description logic knowledge base $\mathcal{K} = (\mathcal{T}, \mathcal{A})$, with $\mathcal{T}$ as given above and $\mathcal{A}$ given as:

$$\mathcal{A} = \{O^\star(a), \exists r.\neg(+!!)(a)\}$$

If this knowledge base is unsatisfiable, we can conclude, that $(+!!)$ will come to pass, if $O^\star$ is operative. Again we can perform this satisfiability test, by translating the TBox and the ABox into DL-clauses and using Hyper to check the satisfiability. We obtain the desired result, namely that (only) ethical code $O^\star$ leads to the most desirable behavior $(+!!)$.

4.3   Experiments

We formalized the examples introduced in this section and tested it with the Hyper theorem prover as described above. Since all formalizations are available

Table 4. Time in seconds Pellet needed to show the unsatisfiability of the introduced examples. Time in seconds Hyper needed to show the unsatisfiability of the DL-clauses for the examples (the second number includes the translation into DL-clauses).

	Multi-agent Systems	Multi-agent Systems (with TBox)	Contrary-to-duty Obligations
Pellet	2.548	2.468	2.31
Hyper	0.048 / 2.596	0.048 / 2.102	0.03 / 1.749

in $\mathcal{ALC}$, we used the description logic reasoner Pellet [20] to show the unsatisfiability of the formalizations as well. Table 4 shows the results of our experiments. In the first column we see the time in seconds the two reasoners needed to show the unsatisfiability of the formalization of the example from multi-agent systems. For Hyper we give two different numbers. The first number is the time Hyper needs to show the unsatisfiability given the set of DL-clauses. In addition to that the second number contains the time needed to transform the $\mathcal{ALC}$ knowledge base into DL-clauses. The second column gives the runtimes for the example from multi-agent systems using the formalization with a TBox. And in the last column we present the runtimes for the consistency test of normative system $\mathcal{N}_1$ from the example on contrary-to-duty obligations.

For the examples we considered, the runtimes of Pellet and Hyper are comparable. Further investigation and comparison with other modal and/or description logic reasoning tools is required and subject of future work. In order to use Hyper to perform the satisfiability tests, we first have to translate the examples into DL-clauses. Our experiments show, that this translation is not harmful.

5   Conclusion

In this paper, we have demonstrated that by means of deontic logic complex normative systems can be formalized easily. These formalizations can be checked effectively with respect to consistency and independence from additional formulae. For normative systems described with deontic logic, there is a one-to-one translation into description logic formulae. These formula can be checked automatically by automated theorem provers, which is in our case Hyper.

We are aware that deontic logic has several limitations. This is why future work aims at using more recent formalisms. For example we want to apply deontic logic in the context of natural-language question-answering systems. There the normative knowledge in large databases often leads to inconsistencies, which motivates us to consider combinations of deontic with defeasible logic.

References

1. Artosi, A., Cattabriga, P., Governatori, G.: Ked: A deontic theorem prover. In: On Legal Application of Logic Programming, ICLP 1994, pp. 60–76 (1994)

2. Baader, F., Nutt, W.: Basic description logics. In: Baader, F., Calvanese, D., McGuinness, D., Nardi, D., Patel-Schneider, P.F. (eds.) The Description Logic Handbook: Theory, Implementation, and Applications, pp. 43–95. Cambridge University Press (2003)
3. Bassiliades, N., Kontopoulos, E., Governatori, G., Antoniou, G.: A modal defeasible reasoner of deontic logic for the semantic web. Int. J. Semant. Web Inf. Syst. 7(1), 18–43 (2011)
4. Baumgartner, P., Furbach, U., Niemelä, I.: Hyper tableaux. In: Alferes, J.J., Pereira, L.M., Orlowska, E. (eds.) JELIA 1996. LNCS, vol. 1126, pp. 1–17. Springer, Heidelberg (1996)
5. Beirlaen, M.: Tolerating normative conflicts in deontic logic. PhD thesis, Ghent University (2012)
6. Bender, M., Pelzer, B., Schon, C.: System description: E-KRHyper 1.4 - Extensions for unique names and description logic. In: Bonacina, M.P. (ed.) CADE 2013. LNCS (LNAI), vol. 7898, pp. 126–134. Springer, Heidelberg (2013)
7. Bringsjord, S., Arkoudas, K., Bello, P.: Toward a general logicist methodology for engineering ethically correct robots. IEEE Intelligent Systems 21(4), 38–44 (2006)
8. Chisolm, R.M.: Contrary-to-duty imperatives and deontic logic. Analysis 23, 33–36 (1963)
9. Furbach, U., Schon, C.: Deontic logic for human reasoning. CoRR, abs/1404.6974 (2014)
10. Gabbay, D., Horty, J., Parent, X., van der Meyden, R., van der Torre, L. (eds.): Handbook of Deontic Logic and Normative Systems. College Publications (2013)
11. Horty, J.F.: Agency and Deontic Logic. Oxford University Press, Oxford (2001)
12. Klarman, S., Gutiérrez-Basulto, V.: Description logics of context. Journal of Logic and Computation (2013)
13. McNamara, P.: Deontic logic. In: Zalta, E.N. (ed.) The Stanford Encyclopedia of Philosophy. Stanford University (2010)
14. McNamara, P., Prakken, H.: Norms, Logics and Information Systems: New Studies in Deontic Logic and Computer Science. Frontiers in artificial intelligence and applications. IOS Press (1999)
15. Motik, B., Shearer, R., Horrocks, I.: Optimized Reasoning in Description Logics using Hypertableaux. In: Pfenning, F. (ed.) CADE 2007. LNCS (LNAI), vol. 4603, pp. 67–83. Springer, Heidelberg (2007)
16. Murakami, Y.: Utilitarian deontic logic. In: Proceedings of the Fifth International Conference on Advances in Modal Logic (AiML 2004), pp. 288–302 (2004)
17. Plaisted, D.A., Greenbaum, S.: A structure-preserving clause form translation. J. Symb. Comput. 2(3), 293–304 (1986)
18. Schild, K.: A correspondence theory for terminological logics: Preliminary report. In: Proc. of IJCAI 1991, pp. 466–471 (1991)
19. Schmidt, R.A., Hustadt, U.: First-order resolution methods for modal logics. In: Voronkov, A., Weidenbach, C. (eds.) Programming Logics. LNCS, vol. 7797, pp. 345–391. Springer, Heidelberg (2013)
20. Sirin, E., Parsia, B., Grau, B.C., Kalyanpur, A., Katz, Y.: Pellet: A practical OWL-DL reasoner. Web Semantics: Science, Services and Agents on the World Wide Web 5(2), 51–53 (2007)
21. von Kutschera, F.: Einführung in die Logik der Normen, Werte und Entscheidungen. Alber (1973)
22. Pelzer, B., Wernhard, C.: System description: E- KRHyper. In: Pfenning, F. (ed.) CADE 2007. LNCS (LNAI), vol. 4603, pp. 508–513. Springer, Heidelberg (2007)

Image Processing Tool for FAE Cloud Dynamics

Mousumi Roy[1], Apparao Allam[2], Arun Agarwal[1],
Rajeev Wankar[1], and Raghavendra Rao Chillarige[1]

[1] SCIS, University of Hyderabad, Gachibowli, Hyderabad-500046, India
[2] High Energy Materials Research Laboratory (HEMRL), DRDO,
Ministry of Defence, Sutarwadi, Pune - 411 021, India

Abstract. Understanding the Fuel air explosive cloud characteristics is
an important activity in FAE warhead designing. This paper develops
and demonstrates the understanding of cloud dynamics through image
processing methodology by analyzing the video. This paper develops apt
ROI extraction method as well as models for cloud radius and height.
This methodology is validated using one of the HEMRL(High Energy
Materials Research Laboratory), Pune experimental data.

Keywords: FAE explosion, Cloud, MBR, cloud ROI.

1 Introduction

FAE warhead is an integrated system used to work against soft targets in a given
region of interest which is designed on two events phenomena. An important
phenomena is cloud formation which is denoted as FAE cloud. The knowledge
about the characteristics of this FAE cloud assist for deriving optimal results.
For understanding the cloud dynamics Dong et.al[3] extracted the cloud images
of FAE explosion and calculated the fractal dimension of the boundary curve
of the cloud and analyzed how the cloud dimension changes with time. The
fractal dimension of solid state FAE explosive images were calculated using box
dimension method[3]. They provided some preliminary foundation which requires
further study on this [3]. As on today, FAE warhead works with open loop
control system because of its complexity. The present study aims at the real
time understanding of the cloud dynamics with the aim to achieve closed loop
control mechanisms.

As the high speed video cameras are a part of the experimentation, it is
proposed to deploy image processing techniques for understanding fuel air cloud
dynamics.

This paper provides methodologies using image processing tools for understanding the cloud dynamics through videos. This paper is organized as follows:
overview of FAE system and a brief description of FAE experimental set up in
section 2. The proposed methodology is developed in section 3. Section 4 provides implementation details. Section 5 deals with experimental results and its
analysis. Conclusions are provided in section 6.

M.N. Murty et al. (Eds.): MIWAI 2014, LNAI 8875, pp. 69–80, 2014.

2   Overview of FAE System and FAE Experimental Set Up

2.1   Overview of FAE System

FAE device consists of a cylindrical fuel container (canister) with an axially fitted explosive charge for the dispersion of the fuel in the form of small droplets which mixes with atmospheric air for the formation of cloud. The cloud is detonated after a predetermined delay for the fuel to attain the explosive limits in the air, with the help of second explosive charge [1]. FAE then produces high impulse blast which has much more severe effect than conventional high explosives [2].

Systems like FAE warhead or integrated systems are highly complex in nature, knowledge related to this type of system is sparse. In understanding FAE system, there are several ways to acquire the data. One of source of such a data are the recordings of the experiment carried out in a form of videos. Treating the video as experimental data and building knowledge through image processing and analysis techniques is not explored. The formation of unconfined fuel and detonation, which depends on the structural aspects of fuel container, property of fuel, means of fuel dispersion and method of initiation are thoroughly studied by Apparao[4]. The theoretical base was developed for the different stages of FAE by conducting the actual experimentations. The unconfined fuel detonates on applying initiator charge when the fuel concentration in air attains the favorable range. The detonable of different height to diameter ratio (H/D) were generated to validate the theoretical estimations which states that cloud with lower height to diameter ratio (H/D) produces higher blast pressures at longer distances, enhancing the lethal area as studied by Rao. C.R. et. al[5]. Most of the real world systems developed based on the image processing methodologies requires ROI (region of interest) extraction at the initial stage. Some of the ROI extraction techniques are detailed as follows : ROI extraction for palm print recognition [7], hand gesture recognition system [8]. A simple and fast region of interest extraction method for identifying the objects in sports scene images is proposed in [9]. ROI extraction in medical imaging are described in [10]. A novel approach for extracting the ROI in infrared images are presented in [11]. [12] describes a novel method for ROI extraction by converting first the color image into gradient map, then using morphological opening and closing operations and finally thresholding the image. One can note that the ROI extraction is a subjective strategy i.e. one need to design ROI extraction methodology case by case.

2.2   FAE Experimental Set Up

A container known as canister with predefined height and diameter filled with liquid fuel is placed on a tripod stand at a specified height. A cylindrical tube with lower diameter (than canister) used as a burster tube filled with burster explosive charge is placed co-axially at the center of the canister. Few serrations are made on the canister wall, here thickness is very less and fuel disperses uniformly in all directions due to these designed serrations. When electronic spark is provided in the burster charge it explodes and the fuel ignites causing the canister to break along its serrations. The liquid fuel disperses in air as small droplets and mixes with atmosphere to form detonable cloud(which is referred as FAE cloud in the rest of the paper). An Initiator for providing second explosive charge to detonate the cloud, is placed at a predefined distance from the center of the canister. There should be some predefined time delay for fuel to attain the concentration level for detonation, after which the electronic spark is provided to Initiator to detonate the cloud and the final blast takes place. The FAE experimental set up is shown in Fig.1 [4].

Two high speed video cameras, namely video camera-1 and video camera-2 which are placed as shown in the Fig.1, captures the video of the explosion. The two videos of FAE, captured by these two different cameras from different angles are referred as Video -1 and Video -2 respectively. In general the duration of the experiment is 300-400 ms.

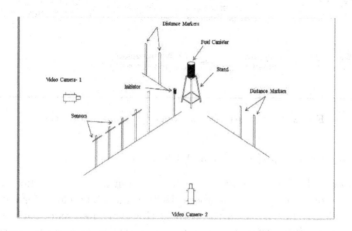

Fig. 1. Experimental set up of FAE

3   Cloud Dynamics: Image Processing Tools (CDIPT)

The cloud formation is a continuous phenomena and is expected to possess cylindrical symmetric about the principal axis of the canister due to the serration on the canister. The video of the FAE cloud formation is nothing but an oval shaped growing region indicating this cloud. Hence it is proposed to extract a contiguous patch of pixels which possess the characteristics (intensity, illumination,...) of cloud for each frame of the video.

As a process, the video is first divided into frames referred to as frame extraction and then transforming it to a gray scale image. Then for each frame, the ROI (region of interest) is extracted by developing the correspondence between the pixel values and cloud characteristics using Image Processing(IP) techniques.

Using minimum bounding rectangle (MBR) and ROI, the radius and the corresponding height of the cloud for that frame over time have been tabulated and empirical models, which have been developed to characterize the cloud dynamics are provided in section 3. The respective derivatives are used to estimate the velocity of the cloud. Figure 2 depicts the flow chart of the cloud velocity estimation process.

The following subsection narrates a schematic methodology of ROI extraction of cloud from each frame.

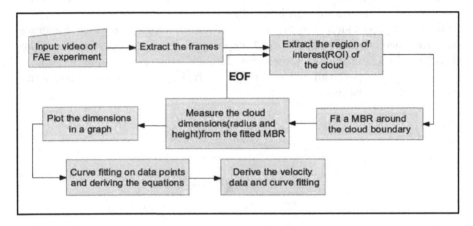

Fig. 2. Flow chart of cloud velocity estimation process

3.1   Frame Extraction from the Video

The video of FAE cloud formation and detonation has been studied frame by frame. This video record has got about 1000 frames. The cloud formation and growth can clearly be visible in the record. However analysis of each frame for growth of cloud is inconvenient and growth is very slow with respect to the time. For getting the better data for correct estimation of cloud growth we have taken frames at every 5 ms. The procedure of frame extraction is as follows:

1. Read the video file and store the structure of the video in some variable.
2. Find the number of frames of the video.
3. Read all the static frames in an array.
4. Store each frame in a specified location with proper extension / format(like .jpg, .png etc.).

The extracted frames of the video are used for further processing as described in next subsection.

3.2   Cloud ROI Extraction

A frame of the video is treated as an image f and its corresponding gray image is denoted by f_g. The image is vertically divided into two parts by considering the principal axis of the canister. The left part of the image called as f_g^L and the corresponding right part is called as f_g^R. By using the pixel threshold(obtained from the thresholding method) the gray images will be converted to binary images by using high pass filter and denoted by f_b^L and f_b^R respectively. Apply median filter to remove noise resulted due to the thresholding. Remove small scattered patches from the image which are non cloud parts from both f_b^L and f_b^R respectively. The fragmented patches of the cloud in the right part of the image f_b^R are joined using run length smoothing algorithm (RLSA). The contiguous patch in the vicinity of the canister is considered as the part of the cloud and the rest of the patches are ignored. The corresponding images are called f_b^{LF} and f_b^{RF}. Combine f_b^{LF} and f_b^{RF} as one image, fill gaps / holes (due to processing) inside the cloud and treat that as region of interest as f_o. Perform morphological operations(closing and opening) to smooth the cloud boundary. The stage wise output of the algorithm is depicted in Fig.3. It is assumed that the total number of frames depicting the stages of the cloud formation are N. These steps are applied on all the N frames.

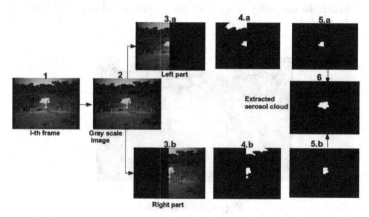

Fig. 3. Pictorial view of the stages of cloud ROI extraction

Issues Related to the Cloud Extraction Process

As FAE experiment is in unconfined environment, the influence of environmental factors will differ from abstraction. To address some of the skew characteristics, cropping up because of the influence of environment and also due to the experimental set up, a schematic method is being proposed. During the cloud extraction process, various aspects have been considered for reproducing the actual cloud image. The cloud image is not a single cloud but it is also showing the pressure gauges in front of the cloud. These issues should be addressed for cloud extraction process.

Methodology Adopted for Addressing the Issues

The proposed methodology considers the image associated with a frame as two images namely left and right with reference to the coaxial of the canister. These two images are subjected to adaptive binarization and smoothening and the fusion of processed left and right parts of the image as shown in figure 3.

Thresholding: It is needed to fix the threshold intensity band denoted as High Pass Filter (HPF) to allow only the cloud pixels ignoring the rest of the background to extract the cloud ROI. The distribution characteristics of the intensity of the cloud pixels and the distribution corresponding to the intensity of the non cloud pixels (background) are considered for selecting the threshold intensity. Accordingly, histograms have been drawn by taking the video frame of the cloud. From the histogram of the cloud pixels and the histogram of non cloud pixels (background), for frames at different time instances, the range of the intensity is selected. Fig.4, 5 and 6 shows the video frame(in gray scale), histogram of the cloud pixels and histogram of background pixels respectively.

It is observed that, selecting the HPF 200-255 some patches of the cloud are lost. Similarly, taking HPF as 90-255, some patches of the background are included into the cloud. Hence various bands of intensity have been tried to find the suitable band to represent the cloud without including the background. During the process it was found that the intensity band 120-255 represents the exact cloud without the overlap of background. Thus the HPF 120-255 has been selected as threshold intensity to extract the optimum cloud ROI from the background.

Fig. 4. Video frame

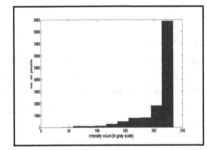

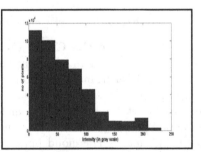

Fig. 5. Histogram of the cloud intensity

Fig. 6. Histogram of the background

3.3   Representing the Cloud ROI by MBR

After identifying the cloud ROI, the next step is to fit a MBR around the cloud ROI denoted as f_o as shown in Fig.7. The axis aligned MBRs of selected video frames of the cloud have been drawn for measuring the cloud dimensions. The method adopted for representing the cloud ROI by MBR is given below:

- Identification of the boundary or contour pixels of the cloud ROI f_o (from ROI extraction algorithm described in previous subsection)and store the pixels coordinate(x,y) in a vector say, P, where 'x' corresponds to the row and 'y' corresponds to the column of the image matrix.
- Determination of x_{min}, x_{max}, y_{min} and y_{max} from the vector P.
- Calculation of the width(w_{mbr}) and height(h_{mbr}) of the MBR as

$$w_{mbr} = x_{max} - x_{min}$$
$$h_{mbr} = y_{max} - y_{min} \tag{1}$$

- Drawing the axis-aligned MBR around the cloud boundary using x_{min}, y_{min}, w_{mbr} and h_{mbr} and store in a new image frame f_i^{mbr}.

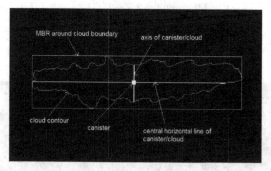

Fig. 7. Cloud contour with fitted MBR

3.4 Radius and Height Determination Process

Radius: The radius ($r_a(t)$) of the cloud is measured as diameter divided by 2, where the diameter is equal to the width of the MBR as shown in Fig.7. It is assumed that the central horizontal line of canister coincides with that of the cloud. First the coordinates (x,y) of the boundary pixels of the cloud ROI f_o are stored in some vector say, P. Then x_{min}, x_{max}, y_{min} and y_{max} are found from the vector P. The radius ($r_a(t)$) of the cloud is measured as

$$D = x_{max} - x_{min}$$
$$r_a(t) = D/2 \tag{2}$$

Height: The height ($h_a(t)$) of the cloud is measured by taking the weighted average of the height measured at different positions of the cloud as shown in figure 8. The boundary informations of the cloud are stored in a vector P. The heights measured as given in the Eqn. 3 at distances 0, $\pm r/3$, $\pm 2r/3$ from the cloud center, by extracting the corresponding boundary pixel informations at these locations from vector P. The weighted average of these heights with preassigned weights gives an estimator for the cloud height $(h_a(t))$. Each of the heights h1, h2, h3, h4 and h5 are measured as follows

$$h1 = y1_{max} - y1_{min}$$
$$h2 = y2_{max} - y2_{min}$$
$$h3 = y3_{max} - y3_{min} \tag{3}$$
$$h4 = y4_{max} - y4_{min}$$
$$h5 = y5_{max} - y5_{min}$$

where, $x1 = x_{center}$, $x2 = x^1_{right} = r + r/3$, $x3 = x^2_{right} = r + 2 * r/3$, $x4 = x^1_{left} = r - r/3$, $x5 = x^2_{left} = r - 2 * r/3$
$y1_{max}, y1_{min}, y2_{max}, y2_{min}, y3_{max}$ and $y3_{min}$ are obtained from vector P. Finally the average height $(h_a(t))$ is calculated as

$$h_a(t) = (h1 * w1 + h2 * w2 + h3 * w3 + h4 * w2 + h5 * w3)/W \tag{4}$$

where $w1 = 5, w2 = 3$ and $w3 = 1$ and $W = w1 + 2 * w2 + 2 * w3$

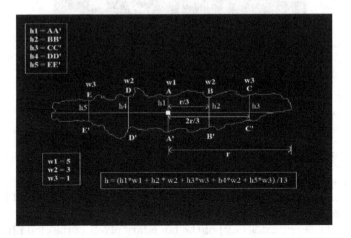

Fig. 8. Pictorial representation of measurement of the cloud height

The radius and the corresponding height of the cloud is measured as described above, for each frame i.e for different time instances and are plotted in a graph as a function of time(t). Due to the non linear nature of the graph, several different curves (like polynomial, exponential, Gaussian etc. and customized) are experimentally fitted along the points in the graph, among which the customized curve resulted the best fit. From the equation of the curve the radius $r_m(t)$ and the corresponding height $h_m(t)$ of the cloud at time 't' are derived as follows

$$r_m(t) = a_r * (1 - e^{(-b_r * t)}) \tag{5}$$

Where $r_m(t)$ = modeled radius of the cloud at time t, a_r and b_r are constants.

$$h_m(t) = a_h * (1 - e^{(-b_h * t)}) \qquad (6)$$

Where $h_m(t)$ = modeled height of the cloud at time t, a_h and b_h are constants. The constants of the above equations a_r, b_r, a_h and b_h are obtained using the non linear least square fitting. From the above modeled equations, the dimensions (radius and height) of the cloud can be measured at any time 't'. The derivatives of the above equations 1 and 2, measures the velocity of the cloud along horizontal $(v_x(t))$ and vertical axes$(v_y(t))$ respectively as follows

$$dr_m(t)/dt = d/dt(a_r * (1 - e^{(-b_r * t)}))$$
$$v_x(t) = a_r * b_r * e^{(-b_r * t)} \qquad (7)$$

Where $v_x(t)$ = velocity of the cloud along the horizontal axis, a_r and b_r are constants.

$$dh_m(t)/dt = d/dt(a_h * (1 - e^{(-b_h * t)}))$$
$$v_y(t) = a_h * b_h * e^{(-b_h * t)} \qquad (8)$$

Where $v_y(t)$ = velocity of the cloud along the vertical axis, a_h and b_h are constants.

4 Implementation

Matlab Software is used along with Windows 7 operating system to implement the above algorithms. Functions similar to standard matlab functions are used for noise removal, morphological operations etc [6].

5 Demonstration and Validation

- A video of FAE experiment conducted by FAE division HEMRL, Pune of 4.2 kg class has been taken for demonstrating the proposed methodology. **Video specifications:** Frame rate is 1ms and cloud data at each 5ms have been used for the studies on cloud dynamics. The size of the extracted frame is 600*800 in pixel unit, the matrix of same size is used to store the image data.
- **Cloud intensity Band:** By applying the thresholding method, the intensity range is fixed at 120-255 in gray scale for extracting the cloud from the background.
- **Pixel Resolution:** From the actual ground distance between the wooden markers, the height of initiator, height and diameter of canister etc. and the pixel distance of teh same, the pixel resolution is, 1 pixel distance = 18mm ground distance(approx) for Video -1.
- Few snapshots of the video and corresponding ROI and MBR obtained through our proposed method are depicted in Fig.9.

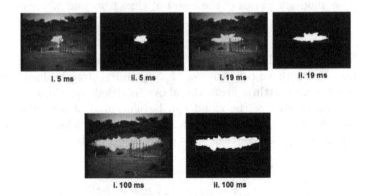

Fig. 9. i. Snapshot of the frames ii. Corresponding ROI and MBR

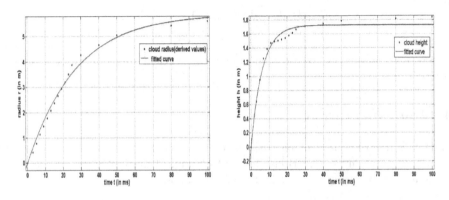

Fig. 10. Plot of the radius of the cloud with respect to time

Fig. 11. Plot of the height of the cloud with respect to time

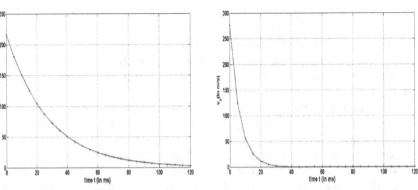

Fig. 12. Plot of the velocity(v_x) along the horizontal axis of the cloud with respect to time

Fig. 13. Plot of the velocity(v_y) along the vertical axis of the cloud with respect to time

The plots of radius and height of the cloud (Video-1) along with the fitted curves are depicted in the Figures 10 and 11 respectively. The corresponding cloud velocities are given in Eqn.7 and 8 respectively and are depicted in Fig.12 and Fig.13 respectively. The value of the constants of the radius and height model are estimated as $a_r = 5.884$, $b_r = 0.0369$, $a_h = 1.728$ and $b_h = 0.1598$ respectively.

5.1   Validation of the Model with the Experimental Data of HEMRL (4.2 kg FAE)

The experimental values of radius of cloud with respect to time has been plotted in Figure 14 and corresponding curve fitting is also carried out in Figure 14. The coefficients of curve have also been derived. The corresponding velocity of the cloud with respect to time has been depicted in Figure 15. The experimental r-t and v-t curves have been compared with that of the corresponding curves derived from the methodology adopted in this work. The radius of the cloud at 100ms and initial velocity obtained from experimental data, as compared with that of the derived data, from image analysis of the video, are presented in table 1. It is observed that there is a very good agreement in the values of radius and initial velocity of the cloud. So it can be concluded that, the image processing approach has got high potential for understanding the cloud dynamics.

Table 1. Comparative data of FAE (4.2 kg) cloud - experimental vs derived values

Srl. no.	Parameters value	From the FAE Experiment(HEMRL)	From Image Analysis
1	radius(r)(in m) at 100ms	5.74	5.52
2	Initial velocity(v_x)(in m/s)	230.89	221

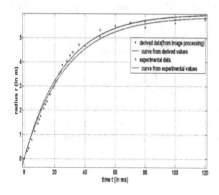

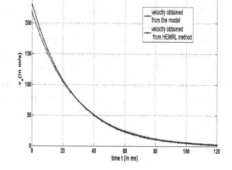

Fig. 14. Cloud radius with respect to time

Fig. 15. Cloud velocity(v_x) with respect to time

6 Conclusion

This paper develops a schematic region of interest extraction method of a gray image corresponding to a frame of experimental video of FAE. An apt cloud height estimation method has been developed. The methodologies have been validated by considering the experimental video from HEMRL. One can extend this study by establishing appropriate correspondence between the cloud characteristics and pixel properties for understanding the other components of the cloud beside the radius, height and velocity.

References

1. Apparao, A.: Fuel Air Explosives. Defence Science Journal 37(1), 23–28 (1987)
2. http://en.wikipedia.org/wiki/Thermobaric_weapon
3. Dong, Y., Li, H., Shi, H., Yi, L.: Study of FAE Explosive Image Analysis Based on the Fractional Dimension. Journal of Mathematics Research 1(1) (March 2009)
4. Apparao, A.: Study on Formation and Detonation of Unconfined Fuel Aerosols and Parameter Optimisation by Mathematical Modeling, PhD Thesis (April 2013)
5. Apparao, A., Rao, C.R.: Performance of Unconfined Detonable Fuel Aerosols of Different Height to Diameter Ratios. Propellants, Explosives, Pyrotechnics (September 19, 2013), doi:10.1002/prep.201300010
6. Gonzalez, R.C., Woods, R.E.: Digital Image Processing. Pearson, First Impression (2009)
7. Kalluri, H.K., Prasad, M.V.N.K., Agarwal, A.: Dynamic ROI Extraction Algorithm for Palmprints. In: Tan, Y., Shi, Y., Ji, Z. (eds.) ICSI 2012, Part II. LNCS, vol. 7332, pp. 217–227. Springer, Heidelberg (2012)
8. Verma, V.K., Wankar, R., Rao, C.R., Agarwal, A.: Hand Gesture Segmentation from Complex Color-Texture Background Image. In: Ramanna, S., Lingras, P., Sombattheera, C., Krishna, A. (eds.) MIWAI 2013. LNCS, vol. 8271, pp. 281–292. Springer, Heidelberg (2013)
9. Xia, Y., Gan, Y., Li, W., Ning, S.: A Simple and Fast Region of Interest Extraction Approach Based on Computer Vision for Sport Scene Images. In: 2nd International Congress on Image and Signal Processing, CISP 2009, October 17-19, pp. 1–4. IEEE (2009)
10. Shao-zhen, Y., Xian-dong, Y.: Medical image retrieval based on extraction of region of interest. In: International Conference on Bioinformatics and Biomedical Engineering, pp. 1–4. IEEE (2010)
11. Erturk, S.: Region of Interest Extraction in Infrared Images Using One-Bit Transform. IEEE Signal Processing Letters, 952–955 (2013)
12. Zhang, X.-H., Zhu, Y.-Y., Fan, Z.-K.: Region of Interest Automatic Extraction for Color Image Based on Mathematical Morphology. In: Ninth IEEE International Conference on Computer and Information Technology, CIT 2009, pp. 113–117. IEEE (2009)

N-gram Based Approach for Opinion Mining of Punjabi Text

Amandeep Kaur and Vishal Gupta

U.I.E.T., Panjab University, Chandigarh, India
vishal@pu.ac.in, amandeepk.cql@gmail.com

Abstract. Opinion mining is the process of analyzing views, attitude or opinions of a writer or a speaker. Research in this particular area involves the detection of opinions from the text of any language. Vast amount of work has been done for the English language. In spite of lack of resources for Indian languages, work has been done for Telugu, Bengali and Hindi language. In this paper, we proposed a hybrid research approach for the emotion/opinion mining of the Punjabi text. Hybrid technique is the combination of Naïve Bayes and N-grams. As the part of presented research, we have extracted the features of N-grams model which are used to train Naïve Bayes. The trained model is then validated using the testing data. Results obtained are also compared with already existing approaches and the accuracy of the results shows the better efficacy of the proposed method.

Keywords: Naïve Bayes, N-grams, Punjabi language.

1 Introduction

Today, Punjabi is widely used and well spoken language in the various parts of world. This language has 100+ million speakers and wide coverage area across the web. But the thing, which is scarce, is the resources and tools to do successive research in this language. In this research, we used some tools of native and other languages to develop our system such as Hindi subjective Lexicon developed by Piyush [7], Punjabi dictionary which contains 35031 words. We have also tested the our system for Punjabi language. The present research has provided following contribution-

- Developed the Punjabi dataset by collecting data from Blogs and Newspapers.
- Generated corpus dataset for training and testing the system.
- Generation of Algorithm for stemming using the rules given by Gupta (2014) [24].

2 Approach Used

Through the literature survey, we come up with the conclusion that machine learning classifier is the basic need for developing a method of Opinion mining. So, as the part of this research, we have used weka module for implementing the classifier named Naïve Bayes. Naïve Bayes technique is used as the machine learning technique in our

M.N. Murty et al. (Eds.): MIWAI 2014, LNAI 8875, pp. 81–88, 2014.
© Springer International Publishing Switzerland 2014

research. This hybrid system is developed by integrating the N-gram model and Naïve Bayes. N-gram model is used for the extraction of features which are further provided for the training of Naïve Bayes. Punjabi Dataset is generated by collecting data from Punjabi newspapers [25] [26] [27] [28] [29] [30] [31] [32] [33] [34] [35] [36] [37] [38] [39] and Blogs [40] [41] [42] [43] [44] [45]. For the implementation of the proposed work, we have chosen the Java programming language. Java language provide us the following services:

- Weka: It is java based toolkit developed at University of Waikato, New Zealand for machine learning.[45]
- Naïve Bayes: This classifier uses Bayes Theorem which relates the probabilities of event given and event already occurred.

2.1   Stemming

Stemming is the process of the generation of stemmed form of a specific token. Table 1 is provided with some root words, which are generated under the process of stemming.

Table 1. Stemming

Word	Root Word	Suffix
ਵੱਡੀ	ਵੱਡਾ	ੀ
ਸੁੰਦਰਤਾ	ਸੁੰਦਰ	ਤਾ
ਕੋੜੀ	ਕੋੜਾ	ੀ

Algorithm for Stemming. We have Formulated the Following Algorithm for Stemming

1. Develop Array of Suffixes which are to be identified and removed to get a Root Word. List of suffixes is collected from research By Dr.Gupta [24]. Say this Array as SuffixArray.
2. Boolean = Check wordtobeStemmed Exists in the Dictionary.
3. If Boolean = true then no Stemming required.
   Else
   RootWord=getRootWord(WordtobeStemmed)
   getRootWord (WordtobeStemmed)
   {
   Position = 2
   Length = length of WordtobeStemmed
   Loop
   Position < length
   Suffix = Split the word at position
   Flag = check if Suffix exits in the SuffixArray developed in Step 1.
   If Flag = true
   Output = ReplaceSuffix (WordtobeStemmed,Suffix)

Return Output
End if
If flag = false and position = length-1
Return WordtobeStemmed // no stemming possible
End if
Position = Position + 1
End loop
}
ReplaceSuffix (WordtobeStemmed , Suffix)
{
Develop two arrays to determine replacement of suffix with kanna (ਾ) or BadiE(ੀ):

KannaArray = { "ੇ","ੋ","ਿਆ","ਿਉ","ਿਉਂ" }

BadiEArray = {"ਿਏ","ਿਉ" }

If Suffix lies in KannaArray
Then Replace the suffix with Kanna (ਾ)

Else if (Suffix lies in BigEArray)
Then Replace the suffix with BadiE (ੀ)

Else
Remove the suffix from WordtobeStemmed
End if
Return Root_word
}

2.2   Negation Handling Process

There are some specific tokens which act as negative words like-"ਨਹੀਂ", "ਨਾ", "ਨਾਮਾਤਰ", "ਨਾਂਹ", "ਨਾਮੁਨਿਕਨ", "ਨਹੀ". The presence of these words invert the polarity from negative to positive and vice versa. So, we have prepared a list of these tokens to handle this issue.

2.3   N-grams

In this technique, Following steps are carried out:

- Corpus is divided into Testing corpus and Training set.
- Using the Training dataset, We developed a N-gram model.
- For each input from Testing set, create the trigrams.
- Match these trigrams with trigrams of training data using the N-gram model
- If matches then increment the value of trimatched (variable name) else create the bi- grams and repeat the previous step for bigrams and unigrams and increment the value of respective match.

Table 2. Feature Set

Class	Features
Positive	No. of unigrams(Positive)
	No. of bigrams(Positive)
	No. of trigrams(Positive)
Negative	No. of unigrams(Negative)
	No. of bigrams(Negative)
	No. of trigrams(Negative)

2.4 Feature Set

Feature Set using Subjective Lexicon and n-grams is constructively shown in Table 2. Increase in the number of features directly affect the increase in the accuracy of system shown in Fig. 1.

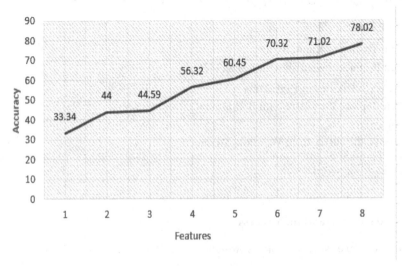

Fig. 1. Increasing Accuracy with Increase in no. of features

2.5 Corpus Generation

Punjabi Dataset is generated by collecting data from Punjabi newspapers [25] [26] [27] [28] [29] [30] [31] [32] [33] [34] [35] [36] [37] [38] [39] and Blogs [40] [41] [42] [43] [44] [45]. Statistics of Resultant Dataset is given in Table 3.

Table 3. Statistics: Punjabi Corpus

	Training	Testing	Newspapers	Blogs
No. of documents	600	284	721	163
No. of sentences	30000	14200	36050	8150
No. of sentences per document	50	50	50	50
No. of words(total)	21000	99400	252350	57050
No. of words per document	359	359	350	350

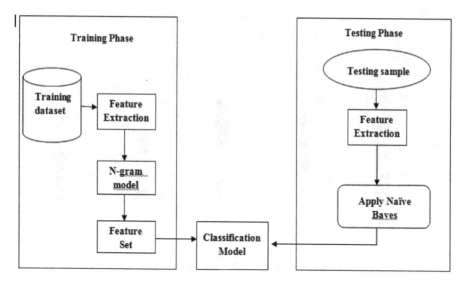

Fig. 2. Process Flow Model

3   Methodology

Process flow of system is described by following Algorithm and Flowchart in Fig.2
Algorithm:

1) Compute the positive trigrams, unigrams, bigrams using N-grams.
2) Compute the negative trigrams, unigrams, bigrams using N-grams.
3) Devise the feature vector which will include the information of step1 and 2. The
   design of Feature Vector is explained in Table 2.This Feature Vector is used for
   Training and Testing Naïve Bayes by using Training and Testing datasets.

4   Result Evaluation

Implementation of system Using Java is done. Input is Punjabi text in Unicode format
or File with encoding equals to UTF-8 can be browsed. Our System takes the input
and provide the result corresponding to Hybrid system and all combination.
Fig 3 shows the results of other approaches used for opinion mining. Hindi Subjective
Lexicon, Hindi Sentiwordnet, Bi-lingual Dictionary, Translated Dictionary. Result
analysis is given in Table 4 and Table 5.

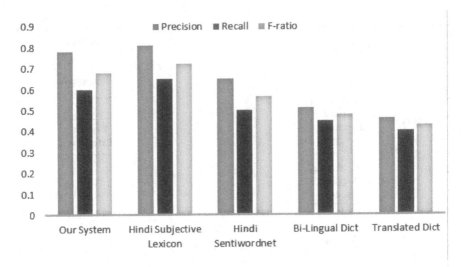

Fig. 3. Comparison with other approaches

Table 4. Result evaluation by accuracy

Technique (Our system)	News	Blogs
N-gram	57.1	56.43
N-gram + negation handling	60.2	63.3
N-gram + negation handling + Stemming	74.1	76.52

Table 5. Parameter Evaluation and Analysis

Technique	Precision	Recall	F-Score
Our System	0.78	0.6	0.67
Hindi Subjective Lexicon	0.81	0.65	0.72
Hindi Sentiwordnet	0.65	0.5	0.56
Bi-Lingual Dictionary	0.51	0.45	0.47
Translated Dict	0.46	0.4	0.42

5   Conclusion and Future Work

Better efficiency shown by our system is great achievement but still required more improvement to get best results. The factors which are responsible for low performance - Lexicon coverage, context dependency, Vocabulary mismatch. Our model basis is deprived of concept of understanding the contextual information. This problem is categorized as contextual dependency. To solve this problem dynamic prior polarity can be added.

This research can be further extended by adding more features like- Punctuation, Word length, Vocabulary Richness, Frequency of function words, Phrase patterns.

References

1. Biadsy, F., Mckeown, K., Agarwal, A.: Contextual phrase-level polarity analysis using lexical affect scoring and syntactic n-grams (2009)
2. Esuli, A., Sebastiani, F., Baccianella, S.: Sentiwordnet 3.0: An enhanced lexical resource for sentiment analysis and opinion mining. In: Proceedings of the Seventh International Conference on Language Resources and Evaluation (LREC 2010), Valletta, Malta (2010)
3. Das, A.: Opinion Extraction and Summarization from Text Documents in Bengali. Doctoral Thesis. UMI Order Number: UMI Order No. GAX95-09398. Jadavpur University (2011)
4. Wiebe, J., Banea, C., Mihalcea, R.: A bootstrapping method for building subjectivity lexicons for languages with scarce resources. In: Proceedings of the Sixth International Language Resources and Evaluation (LREC 2008), Marrakech, Morocco (2008)
5. Bandyopadhyay, S., Das, A.: SentiWordNet for Bangla (2010)
6. Bandyopadhyay, S., Das, A.: SentiWordNet for Indian Languages (2010)
7. Arora, P.: Sentiment Analysis for Hindi Language. Masters thesis, IIT, Hyderabad (2013)
8. Sebastiani, F., Esuli, A.: Sentiwordnet: A publicly available lexical resource for opinion mining. In: Proceedings of the 5th Conference on Language Resources and Evaluation (LREC 2006), p. 54 (2006)
9. McKeown, K.R., Hatzivassiloglou, V.: Predicting the semantic orientation of adjectives. In: Proceedings of the 35th Annual Meeting of the Association for Computational Linguistics and Eighth Conference of the European Chapter of the Association for Computational Linguistics, ACL 1998, Stroudsburg, PA, USA, p. 174 (1997)
10. Liu, B., Hu, M.: Mining and summarizing customer reviews. In: KDD, p. 168 (2004)
11. Wilson, T., Intelligent, T.W.: Annotating opinions in the world press. In: SIGdial 2003, p. 13 (2003)
12. Bhattacharyya, P., Joshi, A., Balamurali, A.R.: A fall-back strategy for sentiment analysis in Hindi: A case study (2010)
13. Rijke, M.D., Kamps, J., Marx, M., Mokken, R.J.: Using wordnet to measure semantic orientation of adjectives. National Institute for, p. 1115 (2004)
14. Indian Institute of Technology, Hyderabad, http://www.iith.ac.in/
15. Kim, S.: Determining the sentiment of opinions. In: Proceedings of COLING, p. 1367 (2004)
16. Hovy, E., Kim, S.: Identifying and analyzing judgment opinions. In: Proceedings of HLT/NAACL 2006, p. 200 (2006)
17. Vaithyanathan, S., Pang, B., Lee, L.: Thumbs up? Sentiment classification using machine learning techniques. In: Proceedings of the 2002 Conference on Empirical Methods in Natural Language Processing (EMNLP), p. 79 (2002)
18. Ravichandran, D., Rao, D.: Semi-supervised polarity lexicon induction. In: Proceedings of the 12th Conference of the European Chapter of the Association for Computational Linguistics, EACL 2009, Stroudsburg, PA, USA, p. 675 (2009)
19. Dunphy, D.C., Stone, P.J., Ogilvie, D.M., Smith, M.S.: The General Inquirer: A Computer Approach to Content Analysis. MIT Press, Cambridge (1966)
20. Turney, P.: Thumbs up or thumbs down? Semantic orientation applied to unsupervised classification of reviews (2002)
21. Wiebe, J.M., O'Hara, T.P., Bruce, R.E.: Development and use of a gold-standard data set for subjectivity classifications. In: Proceedings of the 37th Annual Meeting of the Association for Computational Linguistics on Computational Linguistics, ACL 1999, Stroudsburg, PA, USA, p. 246 (1999)

22. Wilson, T.: Recognizing contextual polarity in phrase-level sentiment analysis. In: Proceedings of HLT-EMNLP, pp. 347–354 (2005)
23. Hatzivassiloglou, V., Yu, H.: Towards answering opinion questions: Separating facts from opinions and identifying the polarity of opinion sentences. In: Proceedings of the 2003 Conference on Empirical Methods in Natural Language Processing, EMNLP 2003, Stroudsburg, PA, USA, p. 129 (2003)
24. Gupta, V.: Automatic Stemming of Words for Punjabi Language. In: Thampi, S.M., Gelbukh, A., Mukhopadhyay, J. (eds.) Advances in Signal Processing and Intelligent Recognition Systems. AISC, vol. 264, pp. 73–84. Springer, Heidelberg (2014)
25. http://www.punjabitribuneonline.com
26. http://www.indotimes.com.au
27. http://www.beta.ajitjalandhar.com
28. http://www.rozanaspokesman.com
29. http://www.dailypunjabtimes.com
30. http://www.deshsewak.in
31. http://www.nawanzamana.in
32. http://www.dailyjanjagriti.com
33. http://www.punjabpost.in
34. http://www.seapunjab.com
35. http://www.ajdiawaaz.com
36. http://www.malwapost.com
37. http://www.punjabinfoline.com
38. http://www.chardhikala.com
39. http://www.deshvideshtimes.com
40. http://www.punjab-screen.blogspot.in
41. http://www.parchanve.wordpress.com
42. http://www.punjabiaarsi.blogspot.in
43. http://www.kamalkang.blogspot.in
44. http://www.shabadsanjh.com
45. http://www.cs.waikato.ac.nz/ml/weka/

Application of Game-Theoretic Rough Sets in Recommender Systems

Nouman Azam and JingTao Yao

Department of Computer Science, University of Regina, Canada S4S 0A2
{azam200n,jtyao}@cs.uregina.ca

Abstract. Recommender systems are emerging as an important business tool in E-commerce. The recommendations in these systems typically rely on some sort of intelligent mechanisms that analyze previous user trends and ratings to make personalized recommendations. In this article, we examine the application of game-theoretic rough set (GTRS) model as an alternative intelligent component for recommender systems. The role of GTRS is examined by considering two important properties of recommendations. The first property is the accuracy of recommendations and the second property is the generality or support of recommendations. It is argued that making highly accurate recommendations for majority of the users is a major hindrance and difficulty for improving the performance of recommender systems. The GTRS meets this challenge by examining a tradeoff solution between the properties of accuracy and generality. Experimental results on *movielen* dataset suggest that the GTRS improves the two properties of recommendations compared to the standard Pawlak rough set model.

Keywords: rough sets, game-theoretic rough sets, recommender systems.

1 Introduction

Recommender systems serve as an aid for guiding the users in decisions on matters related to personal taste and choice [11]. A variety of different techniques and approaches appeared in the literature that are considered as the basis for developing and implementing recommender systems. The most common approaches are content based recommendations, collaborative based recommendations, and demographic based recommendations [3]. Despite of some differences, the core of these approaches is based on utilizing some sort of intelligent mechanisms to make effective recommendations. The collaborative based approach is relatively more popular and successful for building recommender systems [11] and is therefore the focus of this research.

Collaborative recommendations are based on the likings or ratings of users having similar preferences [10]. An understanding in such recommendations is that the users having similar ratings in the past are likely to have similar tastes in the future [10]. However, there are situations where the users having similar

M.N. Murty et al. (Eds.): MIWAI 2014, LNAI 8875, pp. 89–100, 2014.
© Springer International Publishing Switzerland 2014

previous ratings may differ on ratings of unknown or new items. Recommendations in such cases can be very challenging. Relatively accurate recommendations may be possible when similar users (identified based on the past ratings) exhibit a high degree of consistency in their opinions about unknown items. Such recommendations, however, may only be made for a small number of users. Decreasing the requirement of being highly accurate may allow us to make recommendations for more users. In general, recommendations with high accuracy may have lesser support or generality while recommendations with high generality may suffer from lower accuracy [4]. Herlocker et al. [6] referred to these two conflicting dimensions as the strength and support of recommendations and Lin et al. [8] referred to them as confidence and coverage of recommendations. Despite difference in terminologies and possibly to a certain extend in interpretation, they essentially referred to the same properties as accuracy and generality. A tradeoff level needs to be determined between these two properties in order to obtain effective recommendations [4]. We employ the game-theoretic rough set model (GTRS) for investigating such a tradeoff.

A main contribution of GTRS is that it provides an intelligent mechanism for determining thresholds by realizing a tradeoff between multiple cooperative or competing criteria in the probabilistic rough set model [2],[5],[12]. The computed thresholds can be used to obtain the three rough set regions which are helpful in applications for obtaining useful rules for decision support [5],[12]. This article extends the application of GTRS to obtain recommendation decisions. In particular, we focus on determining a tradeoff solution between the properties of accuracy and generality of rough set based recommendations. Experimental results with the GTRS are examined and analyzed on the *movielen* dataset.

2   Overview of Game-Theoretic Rough Set Model

A main result of probabilistic rough sets is that the lower and upper approximations for a concept C are defined using a pair of thresholds (α, β) [13],

$$\underline{apr}_{(\alpha,\beta)}(C) = \bigcup\{x \in U \mid P(C|[x]) \geq \alpha\} \qquad (1)$$

$$\overline{apr}_{(\alpha,\beta)}(C) = \bigcup\{x \in U \mid P(C|[x]) > \beta\} \qquad (2)$$

where U is the set of objects called universe and $E \subseteq U \times U$ is an equivalence relation on U. The equivalence class of E containing object $x \in U$ is denoted as $[x]$ and the concept $C \subseteq U$. The three rough set regions based on lower and upper approximations are defined as,

$$\begin{aligned} \mathrm{POS}_{(\alpha,\beta)}(C) &= \underline{apr}_{(\alpha,\beta)}(C) \\ &= \{x \in U | P(C|[x]) \geq \alpha\}, \qquad (3) \\ \mathrm{NEG}_{(\alpha,\beta)}(C) &= (\overline{apr}_{(\alpha,\beta)}(C))^c \\ &= \{x \in U | P(C|[x]) \leq \beta\}, \qquad (4) \end{aligned}$$

$$\mathrm{BND}_{(\alpha,\beta)}(C) = \overline{apr}_{(\alpha,\beta)}(C) - \underline{apr}_{(\alpha,\beta)}(C)$$
$$= \{x \in U | \beta < P(C||[x]) < \alpha\}, \tag{5}$$

where $P(C||[x])$ is the conditional probability of an object x to be in C given that the same object is in $[x]$. In contrast to the conventional Pawlak model, where $P(C||[x])$ has to be strictly one or zero for an object x to be included in either positive or negative regions, the probabilistic approach introduces (α, β) thresholds to relax these strict membership conditions. Particularly, an object x is considered to be in positive region, if its probabilistic relationship with C is at or above level α, i.e., $P(C||[x]) \geq \alpha$ and in the negative region if its probabilistic relationship with C is at or below level β, i.e., $P(C||[x]) \leq \beta$. In situations, where with the current information, whether or not the object x to be in C can not be determined, if the probabilistic relationship of x with C is between the two thresholds, i.e., $\beta < P(C||[x]) < \alpha$. A major issue in the probabilistic rough sets research is the determination and interpretation of thresholds (α, β) [13]. The GTRS meet these challenges by implementing a game for determining a balanced solution based on multiple factors that leads to cost effective thresholds [5],[12].

A typical game is defined as a tuple $\{P, S, u\}$ where, P is a finite set of n players, $S = S_1 \times ... \times S_n$, with each S_i representing a finite set of strategies for player i and each $(s_1, s_2, ..., s_n) \in S$ is a strategy profile where each player i plays s_i, Finally, $u = (u_1, ..., u_n)$, with each $u_i : S \longmapsto \Re$ representing a real-valued utility or payoff function for player i [7]. The players in GTRS are considered as different criteria highlighting various aspects of rough set based decision making, such as, accuracy or applicability of decision rules [2],[5]. Suitable measures are selected and used to evaluate these criteria. The strategies are realized in term of different or varying levels of the properties, such as the risks, costs or uncertainty associated with different regions which leads to different threshold levels [5],[12]. The strategies can alternatively be formulated as direct modification of thresholds [2]. Each criterion is affected in a different way based on different strategies within the game. The ultimate goal of the game is to find an acceptable solution based on the considered criteria which are used to determine effective thresholds.

Table 1 represents a typical two-player game in GTRS. The players in this game are represented as criteria c_1 and c_2, respectively. Each cell of the table contains a pair of utility functions which is calculated based on the respective strategy profile.

Table 1. Payoff table for a two-player GTRS based game

		c_2		
		s_1	s_2	...
c_1	s_1	$u_{c_1}(s_1, s_1), u_{c_2}(s_1, s_1)$	$u_{c_1}(s_1, s_2), u_{c_2}(s_1, s_2)$	...
	s_2	$u_{c_1}(s_2, s_1), u_{c_2}(s_2, s_1)$	$u_{c_1}(s_2, s_2), u_{c_2}(s_2, s_2)$	...
	...	...	...	...

Each cell of the table corresponds to a strategy profile and contains a pair of utility functions. For instance, the top right cell corresponds to a strategy profile (s_1, s_1) which contains utility functions $u_{c_1}(s_1, s_1)$ and $u_{c_2}(s_1, s_1)$. The game solution, such as the Nash equilibrium, is utilized to determine a possible strategy profile and the associated threshold pair [7]. The determined thresholds are then used in the probabilistic rough set model to obtain three regions and the implied three-way decisions.

3   Applying GTRS in Collaborative Recommendations

We examine two aspects or properties of collaborative recommendations that play a vital role in evaluating and determining the performance of recommendations.

3.1   Properties of Accuracy and Generality of Recommendations

The property of accuracy measures how close a recommender system recommendations are to the actual user preferences. Generally speaking, from the early recommender systems to date, the majority of the published work remained focused on different ways of measuring this property to evaluate recommender systems [9]. Several other evaluation properties beyond the accuracy including the generality or coverage are being proposed and discussed in [6]. The property of generality is interpreted and defined in different ways [6]. In this article, it is considered as the relative number of users for whom we actually make recommendations. The two properties provide different and complimentary aspects for evaluating recommendations using rough sets. We highlight this by considering an example.

Table 2 represents the ratings of users on different movies. A positive sign, i.e., + indicates that the user has liked the movie and a negative sign, i.e., - indicates that the user did not like the movie. Each row in the table represents the ratings of a particular user corresponding to the four movies. Suppose we are interesting in predicting the users ratings on Movie 4 using rough set analysis. One can not only use previous ratings on Movie 4 to predict the rating of a new user on Movie 4. This may be possible when we have additional features of the users or movies such as demographic information or user ratings on other movies. We make use of the rationale of collaborative recommendations, i.e., users having similar taste or preferences over the seen movies are likely to have similar preferences for the unseen movies. This suggests that the similarities between users on ratings of Movie 1, 2 and 3 can be exploited and used to predict the users ratings on Movie 4.

An information table is required to commence with the application of rough sets. Table 2 is essentially an information table since the set of objects (perceived as users) are described by a set of attributes (user ratings for movies). This means that we are able to directly apply rough set analysis on Table 2. Let X_i represents an equivalence class which is the set of users having the same ratings for Movie

Table 2. User ratings for four movies

	Movie 1	Movie 2	Movie 3	Movie 4
U_1	+	+	+	+
U_2	+	+	-	+
U_3	+	-	+	+
U_4	-	+	+	+
U_5	+	+	-	+
U_6	-	+	+	-
U_7	-	+	-	+
U_8	-	-	+	+
U_9	-	+	+	+
U_{10}	+	-	+	+
U_{11}	-	+	+	+
U_{12}	+	-	-	+
U_{13}	+	-	-	-
U_{14}	-	-	-	-
U_{15}	+	-	+	+
U_{16}	+	-	+	+
U_{17}	-	+	-	+
U_{18}	+	-	-	-
U_{19}	-	-	-	-
U_{20}	+	-	+	-
U_{21}	-	+	-	-
U_{22}	-	-	+	-
U_{23}	-	-	+	-
U_{24}	+	-	-	-
U_{25}	-	-	-	-
U_{26}	+	-	-	-

1, 2 and 3. The following eight equivalence classes can be formed based on the data in Table 2,

$$X_1 = \{U_1\}, \qquad\qquad X_2 = \{U_2, U_5\},$$
$$X_3 = \{U_3, U_{10}, U_{15}, U_{16}, U_{20}\}, \quad X_4 = \{U_4, U_6, U_9, U_{11}\},$$
$$X_5 = \{U_7, U_{17}, U_{21}\}, \qquad\qquad X_6 = \{U_8, U_{22}, U_{23}\},$$
$$X_7 = \{U_{12}, U_{13}, U_{18}, U_{24}, U_{26}\}, \quad X_8 = \{U_{14}, U_{19}, U_{25}\}.$$

The concept of interest in this case is to determine the positive ratings for the Movie 4, i.e., Movie 4 = +. We approximate this concept in the probabilistic rough sets framework using Equations (3) - (5). The association of each equivalence class X_i with the concept, i.e., $P(C|X_i)$ needs to be determined in order to obtain the three regions. The conditional probability is given by,

$$P(C|X_i) = P(\text{Movie 4} = +|X_i) = \frac{|\text{Movie 4} = + \cap X_i|}{|X_i|}. \tag{6}$$

The conditional probabilities of equivalence classes $X_1, ..., X_8$ based on Equation (6) are calculated as 1.0, 1.0, 0.8, 0.75, 0.67, 0.33, 0.2 and 0.0, respectively. The conditional probabilities represent the level of agreement between similar users to positively rate the movie. Generally speaking, the users having similar opinions or preferences on certain items, does not necessarily imply that they will always be in perfect agreement on other items. The probability of an equivalence class X_i is determined as $P(X_i) = |X_i|/|U|$ which means that the probability of X_1 is $|X_1|/|U| = 1/26 = 0.038$. The probabilities of other equivalence classes $X_2, ..., X_8$ are similarly calculated as 0.077, 0.192, 0.154, 0.115, 0.115, 0.192 and 0.115, respectively.

Let us look at the properties of accuracy and generality. For a group containing both positive and negative regions we define these measures as [2],

$$Accuracy(\alpha, \beta) = \frac{|(POS_{(\alpha,\beta)} \cap C) \bigcup (NEG_{(\alpha,\beta)} \cap C^c)|}{|POS_{(\alpha,\beta)} \bigcup NEG_{(\alpha,\beta)}|}, \tag{7}$$

$$Generality(\alpha, \beta) = \frac{|POS_{(\alpha,\beta)} \bigcup NEG_{(\alpha,\beta)}|}{|U|}. \tag{8}$$

The three regions in the conventional Pawlak rough set model are determined as $POS_{(1,0)}(C) = \bigcup \{X_1, X_2\}$, $BND_{(1,0)}(C) = \bigcup \{X_3, X_4, X_5, X_6, X_7\}$, and $NEG_{(1,0)}(C) = \{X_8\}$. The properties of accuracy and generality for the Pawlak mode are

$$Accuracy(\alpha, \beta) = \frac{|((X_1 \bigcup X_2) \cap C) \bigcup (X_8 \cap C^c)|}{|X_1 \bigcup X_2 \bigcup X_8|},$$

$$= \frac{|\{U_1, U_2, U_5, U_{14}, U_{19}, U_{25}\}|}{|\{U_1, U_2, U_5, U_{14}, U_{19}, U_{25}\}|} = \frac{6}{6} = 1.0, \tag{9}$$

$$Generality(\alpha, \beta) = \frac{|(X_1 \bigcup X_2 \bigcup X_8)|}{|U|},$$

$$= \frac{\{U_1, U_2, U_5, U_{14}, U_{19}, U_{25}\}}{|U_1, U_2, ..., U_{27}|} = \frac{6}{26} = 0.2307. \tag{10}$$

Equations (9) - (10) suggest that with the Pawlak model, we make recommendations that are 100% accurate, however these recommendation are possible for only 23.07% of the users. Making recommendations for more users may be possible when we lower our expectation of being 100% definite or correct in all the cases. For instance, if we lower our expectation of acceptance for recommendation by lowering the value of α to 0.8, then X_3 will also be included in the positive region. This means that for the objects in X_3, i.e., $\{U_3, U_{10}, U_{15}, U_{16}, U_{20}\}$, we predict positive ratings. However, from Table 2 it is noted that 4 out of 5 of these recommendations will be correct, i.e., U_{20} has a - rating for Movie 4. How much to increase the level of generality at the cost of decrease in the level of accuracy is an important question in this context. This may be approached from the viewpoint of tradeoff between the two properties.

Table 3. The payoff table for the example game

		G		
		$s_1 = \alpha\!\downarrow$	$s_2 = \beta\!\uparrow$	$s_3 = \alpha\!\downarrow\beta\!\uparrow$
	$s_1 = \alpha\!\downarrow$	(0.83,0.69)	(0.85,0.77)	(0.83,0.89)
A	$s_2 = \beta\!\uparrow$	(0.85,0.77)	(0.86,0.54)	**(0.83,0.89)**
	$s_3 = \alpha\!\downarrow\beta\!\uparrow$	(0.83,0.89)	(0.83,0.89)	(0.8077,0.1.0)

3.2   An Approach for Effective Recommendations with GTRS

In the above example, it may be noted that the probabilistic thresholds (α, β) controls the tradeoff between the properties of accuracy and generality. Determining effective values for the (α, β) thresholds would lead to a moderate and cost effective levels for accuracy and generality. We consider a GTRS based approach for this purpose.

Consider a two-player game in GTRS between the properties of accuracy and generality. Each of these players can choose from three possible strategies, namely, strategy $s_1 = \alpha\!\downarrow$ = decrease α, $s_2 = \beta\!\uparrow$ = increase β and $s_3 = \alpha\!\downarrow\beta\!\uparrow$ = decrease α and increase β. We consider an increase or decrease of 25% in this example. Moreover, an initial threshold pair of $(\alpha, \beta) = (1, 0)$ is considered. A particular strategy, say s_1 is interpreted as 25% decrease in α which leads to $\alpha = 0.75$. The thresholds corresponding to a strategy profile will be determined as the sum of the two changes. In case where only one player suggest a change in the thresholds, the threshold value will be determined as an increase or decrease suggested by that player. A threshold pair corresponding to a strategy profile say (s_1, s_2) = (25% decrease in α, 25% increase in β) will be determined as $(\alpha, \beta) = (0.75, 0.25)$. The corresponding utilities of the two players based on the strategy profile, say (s_1, s_2), are determined as the values of $Accuracy(\alpha, \beta)$ and $Generality(\alpha, \beta)$ based on Equations (7) - (8) with $(\alpha, \beta) = (0.75, 0.25)$.

Table 3 represents the payoff table corresponding to this game. The cell containing bold values, i.e., **(0.83,0.89)** with its corresponding strategy profile (s_2, s_3) is the game solution based on the Nash equilibrium [7]. This means that none of the two players can achieve a higher payoff, given the other player's chosen action. The threshold pair based on the game outcome is given by $(\alpha, \beta) = (0.75, 0.5)$. The GTRS based results for Movie 4 are interpreted as follows. We are able to make 83% correct predictions for 89% of the users when we reduce and set the levels for acceptance and rejection of recommendations for the movie at 0.75 and 0.5, receptively. Comparing these results with the Pawlak model, the GTRS provide recommendations for an additional 62% of the users at a cost of only 17% decrease in accuracy. Please be noted that the (accuracy, generality) of the Pawlak model are (100%, 23.07%) as discussed in Section 3.1.

3.3   Repetitive Threshold Learning with GTRS

To examine and evaluate the GTRS performance for predicting the ratings in a recommender system, we consider a similar game as discussed above in Section 3.2. The game is being played between accuracy and generality to obtain the threshold parameters. To calculate a change in a threshold based on a certain strategy (i.e., either $s_1 = \downarrow\alpha$, $s_2 = \uparrow\beta$ or $s_3 = \downarrow\alpha\uparrow\beta$), four types of changes are noted in the payoff table corresponding to the game in Table 3, i.e.,

$$\alpha^- = \text{single player suggests to decrease } \alpha, \tag{11}$$

$$\alpha^{--} = \text{both the player suggest to decrease } \alpha, \tag{12}$$

$$\beta^+ = \text{single player suggests to increase } \beta, \tag{13}$$

$$\beta^{++} = \text{both the player suggest to increase } \beta. \tag{14}$$

A threshold pair during a game corresponding to a strategy profile $(s_1, s_1) = (\alpha_\downarrow, \alpha_\downarrow)$ is determined as (α^{--}, β), since both the player suggest to decrease threshold α (see Equation (12)). Please be noted that we considered the strategies as decreasing levels for threshold α and increasing levels for threshold β since we consider the thresholds corresponding to the Pawlak model as the initial thresholds, i.e., $(\alpha, \beta) = (1, 0)$. This helps facilitate in comparing our results with the Pawlak model.

Modifying the thresholds continuously to improve the utility levels of the players will lead to a learning mechanism. The learning principle employed in such a mechanism is based on the relationship between modifications in thresholds and their impact on the utilities of players. We exploit this relationship in order to define the variables $(\alpha^-, \alpha^{--}, \beta^+, \beta^{++})$. This will help in continuously improving the thresholds to reach their effective values. A repeated or iterative game is considered to achieve this.

Let (α, β) represent the initial threshold values for a particular iteration of the game. The game will use equilibrium analysis to determine the output threshold pair, say (α', β'). Based on the thresholds (α, β) and (α', β'), we define the four variables that appeared in Equations (11) - (14) as,

$$\alpha^- = \alpha - (\alpha \times (Generality(\alpha', \beta') - Generality(\alpha, \beta))), \tag{15}$$

$$\alpha^{--} = \alpha - c(\alpha \times (Generality(\alpha', \beta') - Generality(\alpha, \beta))), \tag{16}$$

$$\beta^{++} = \beta - (\beta \times (Generality(\alpha', \beta') - Generality(\alpha, \beta))), \tag{17}$$

$$\beta^{++} = \beta - c(\beta \times (Generality(\alpha', \beta') - Generality(\alpha, \beta))). \tag{18}$$

The threshold values for the next iteration are updated to (α', β'). The constant c in Equations (15) and (18) is introduced to reflect the desired level of change in thresholds and should be greater than 1 [2]. The iterative process stops when either the boundary region becomes empty, or the positive region size exceeds the prior probability of the concept C, or $Generality(\alpha, \beta)$ exceeds $Accuracy(\alpha, \beta)$.

4   Experimental Results and Discussion

We investigate the usage of GTRS in movie recommendations. The *movielen* dataset is typically considered for movie recommendation application [1]. The version 1M of the *movielen* which contains about 1 million user ratings is considered in this paper. The dataset consists of three different tables, namely, the user table, the ratings table and the movie table. The user table contains information of 6,040 users including their ages, genders and occupations. The movie table contains information of 3,952 movies including their titles and genres. Finally, the ratings table contains 1 million user ratings on a 5-star scale.

For ease in computation, we reduced and considered the ratings of about 400 users which resulted in about 58,000 user ratings. In addition, we converted the 5-star scale to a binary scale ("like", 'dislike"). Two sets of experiments were conducted based on this conversion. In the first setup, the ratings of 4 or 5 indicated "like" while the ratings of 1 to 3 indicated "dislike". In the second setup only the of rating 5 indicated "like" while the remaining indicated "dislike". Moreover, we reduced the number of movies to the top 10 most frequently rated movies by the users. The prediction on each movie was considered based on the user ratings for the other nine movies. For testing the results, 10 fold cross validation was used in all the experiments.

We first look at the results for the first experimental setup. Table 4 presents the results on the training data. The Pawlak model achieves 100% accurate predictions for user ratings. However, these predictions are possible for a certain portion of the users, for instance, in case of Movie 1, these predictions are possible for 64.81% of the users. Arguably, one would like to extend these predictions to cover more users. The GTRS makes it possible by allowing a slight decrease in accuracy. For instance, in case of Movie 1, the GTRS provides an increase in the coverage of predictions by 30.59% at a cost of 8.6% decrease in accuracy. Let

Table 4. Train results for data with rating of 4 or 5 = "like" and 1 to 3 = "dislike"

Prediction for Movie	Accuracy		Generality	
	GTRS	Pawlak	GTRS	Pawlak
1.	0.9140	1.0	0.9540	0.6481
2.	0.9525	1.0	0.9198	0.6878
3.	0.9829	1.0	0.9782	0.8584
4.	0.9712	1.0	0.9546	0.8037
5.	0.9605	1.0	0.9594	0.7798
6.	0.9739	1.0	0.9271	0.7916
7.	0.9792	1.0	0.9875	0.9271
8.	0.9615	1.0	0.9314	0.7512
9.	0.9766	1.0	0.9625	0.8633
10.	0.9687	1.0	0.9825	0.8641
Average	0.9641	**1.0**	**0.9557**	0.7975

Table 5. Test results for data with rating of 4 or 5 = "like" and 1 to 3 = "dislike"

Prediction for Movie	Accuracy		Generality	
	GTRS	Pawlak	GTRS	Pawlak
1.	0.4898	0.4448	0.8965	0.6847
2.	0.6425	0.6426	0.8977	0.7371
3.	0.5873	0.5484	0.9713	0.8730
4.	0.5950	0.5749	0.9537	0.8256
5.	0.5802	0.5659	0.9143	0.8083
6.	0.6348	0.6126	0.9016	0.8145
7.	0.6680	0.6598	0.9838	0.9627
8.	0.6407	0.6303	0.9102	0.7807
9.	0.6194	0.6269	0.9428	0.8827
10.	0.7252	0.7344	0.9750	0.9102
Average	**0.6183**	0.6041	**0.9347**	0.8279

us look at the average of these results for the two models over the 10 prediction problems. The Pawlak model provides 100% accurate predictions for 85.88% of the users. The GTRS provides 95.86% accurate predictions for 96.49% users. This means that on average, decreasing the accuracy by 4.14%, we are able to make recommendations for additional 10.61% users.

The results on testing data are presented in Table 5. The GTRS provides better accuracy for Movie 1, 3, 4, 5, 6, 7 and 8. This means that in 7 out of 10 cases, the GTRS is better than the Pawlak model. The GTRS provides superior results based on the generality of the predictions in all the cases. Considering the average of these results, the GTRS achieves an accuracy of 61.65% compared to an accuracy of 60.41% achieved with its counterpart. The generality of GTRS is 94.5% which is 11.7% higher than the generality with the Pawlakmodel. This suggests that on average the GTRS not only improves the accuracy but also provides improvement in generality.

Let us now look at the results when only the rating of 5 is considered as "like". Tables 6 presents the training results which are similar to the training results in Table 4. There are differences in the testing results, presented in Table 7, compared to the testing results presented in Table 5. Overall we note an increase in the accuracy of the two models. The GTRS outperforms the Pawlak model in all the aspects. The average results further substantiate this. An average accuracy of 74.21% is determined which is 2.13% higher than the Pawlak. The generality of GTRS is 95.89% compared to 88.16%.

The above experimental results demonstrate and advocate the usefulness of GTRS for improving the quality of recommendation. The GTRS is therefore suggested as an interesting alternative for obtaining rough set based recommendations in recommender systems.

Table 6. Train results for data with rating of 5 = "like" and 1 to 4 = "dislike"

Prediction for Movie	Accuracy		Generality	
	GTRS	Pawlak	GTRS	Pawlak
1.	0.9375	1.0	0.9678	0.7529
2.	0.9818	1.0	0.9346	0.8248
3.	0.9807	1.0	0.9548	0.8397
4.	0.9795	1.0	0.9596	0.8251
5.	0.9763	1.0	0.9836	0.8829
6.	0.9799	1.0	0.9522	0.8430
7.	0.9865	1.0	0.9951	0.9572
8.	0.9756	1.0	0.9519	0.8493
9.	0.9828	1.0	0.9750	0.9008
10.	0.9782	1.0	1.0	0.9129
Average	0.9759	**1.0**	**0.9675**	0.8589

Table 7. Test results for data with rating of 5 = "like" and 1 to 4 = "dislike"

Prediction for Movie	Accuracy		Generality	
	GTRS	Pawlak	GTRS	Pawlak
1.	0.6488	0.5962	0.9428	0.7835
2.	0.7613	0.7382	0.9303	0.8418
3.	0.7372	0.7147	0.9526	0.8617
4.	0.6958	0.6631	0.9451	0.8444
5.	0.7092	0.6877	0.9701	0.9017
6.	0.7684	0.7521	0.9477	0.8706
7.	0.7596	0.7554	0.9875	0.9726
8.	0.7676	0.7444	0.9452	0.8655
9.	0.7535	0.7436	0.9775	0.9327
10.	0.8201	0.8131	0.9900	0.9415
Average	**0.7421**	0.7208	**0.9589**	0.8816

5 Conclusion

Recommender systems are gaining significant attention as in E-commerce and E-bussiness. In this article, we examine the application of game-theoretic rough sets or GTRS in the capacity of an intelligent component for making collaborative recommendations in recommender systems. The properties of accuracy and generality are being examined with GTRS which provides useful and in some sense complimentary aspects of recommendations. In an ideal situation, the recommender systems are expected to provide highly accurate recommendations for majority of its users. This however is not always possible and one has to find a suitable level of tradeoff between the two properties. This tradeoff is controlled by a pair of thresholds (α, β) in the probabilistic rough set framework. We investigated the use of GTRS in determining suitable thresholds based on a tradeoff solution between accuracy and generality. Experimental results on *movielen* dataset suggest that GTRS not only provides more accurate recommendations but at the same time significantly improve the generality (between 7-17%) compared to the standard Pawlak model.

Acknowledgements. This work was partially supported by a Discovery Grant from NSERC Canada and the University of Regina Verna Martin Memorial Scholarship Program.

References

[1] Grouplens research, http://www.grouplens.org
[2] Azam, N., Yao, J.T.: Analyzing uncertainties of probabilistic rough set regions with game-theoretic rough sets. International Journal of Approximate Reasoning 55(1), 142–155 (2014)
[3] Bobadilla, J., Ortega, F., Hernando, A., Gutiérrez, A.: Recommender systems survey. Knowledge-Based Systems 46, 109–132 (2013)
[4] Ge, M., Delgado-Battenfeld, C., Jannach, D.: Beyond accuracy: evaluating recommender systems by coverage and serendipity. In: Proceedings of the Fourth ACM Conference on Recommender Systems, pp. 257–260 (2010)
[5] Herbert, J.P., Yao, J.T.: Game-theoretic rough sets. Fundamenta Informaticae 108(3-4), 267–286 (2011)
[6] Herlocker, J.L., Konstan, J.A., Terveen, L.G., Riedl, J.T.: Evaluating collaborative filtering recommender systems. ACM Transactions on Information Systems 22(1), 5–53 (2004)
[7] Leyton-Brown, K., Shoham, Y.: Essentials of Game Theory: A Concise Multidisciplinary Introduction. Morgan & Claypool Publishers (2008)
[8] Lin, W., Alvarez, S.A., Ruiz, C.: Efficient adaptive-support association rule mining for recommender systems. Data Mining and Knowledge Discovery 6(1), 83–105 (2002)
[9] Hernández del Olmo, F., Gaudioso, E.: Evaluation of recommender systems: A new approach. Expert Systems with Applications 35(3), 790–804 (2008)
[10] Pazzani, M.J.: A framework for collaborative, content-based and demographic filtering. Artificial Intelligence Review 13(5-6), 393–408 (1999)
[11] Su, X.Y., Khoshgoftaar, T.M.: A survey of collaborative filtering techniques. Advances in Artificial Intelligence 2009, 1–19 (2009)
[12] Yao, J.T., Herbert, J.P.: A game-theoretic perspective on rough set analysis. Journal of Chongqing University of Posts and Telecommunications (Natural Science Edition) 20(3), 291–298 (2008)
[13] Yao, Y.Y.: Two semantic issues in a probabilistic rough set model. Fundamenta Informaticae 108(3-4), 249–265 (2011)

RGB - Based Color Texture Image Classification Using Anisotropic Diffusion and LDBP

Prakash S. Hiremath and Rohini A. Bhusnurmath*

Department of P.G. Studies and Research in Computer Science,
Gulbarga University, Gulbarga, 585106, India
hiremathps53@yahoo.com,rohiniabmath@gmail.com

Abstract. In this paper, a novel color texture image classification based on RGB color space using anisotropic diffusion, and local directional binary patterns (LDBP) is introduced. Traditionally, RGB color space is widely used in digital images and hardware. RGB color space is applied to obtain more accurate color statistics for extracting features. According to characteristic of anisotropic diffusion, image is decomposed into cartoon approximation; further the texture approximation is obtained by subtracting the original image and cartoon approximation. Then, texture features of image are obtained by applying LDBP co-occurrence matrix parameters on texture approximation. LDA is used to enhance the class seperability. After feature extraction, k-NN classifier is used to classify texture classes by the extracted features. The proposed method is evaluated on Oulu database. Experimental results demonstrate the proposed method is better and more correct than RGB based color texture image classification methods in the literature.

Keywords: RGB color space, texture classification, anisotropic diffusion, LDBP, co-occurrence matrix, k-NN.

1   Introduction

Texture analysis is very useful for experiments of image classification and identification. Thus, it has long been an area of computer vision with active research area spanning image processing, pattern recognition, and computer vision, with applications to medical image analysis, remote sensing, object recognition, industrial surface inspection, document segmentation and content based image retrieval. Texture classification has received significant attention with many proposed approaches, as documented in comprehensive surveys [1], [2], [3], [4], [5]. The ability of a human to distinguish different textures is apparent, therefore, the automated description and recognition of the texture images is in demand. Over the years, many researchers have studied different texture analysis methods. Many of these methods represent the local behavior of the texture via statistical [6], structural [7] or spectral [8] properties of the

---

* Corresponding author.

M.N. Murty et al. (Eds.): MIWAI 2014, LNAI 8875, pp. 101–111, 2014.

image. A methodology is presented in [9], where second order probability distributions [2], [4] are enough for human discrimination of two texture patterns, has motivated the use of statistical approaches. On the other hand, structural approaches describe the textures by rules, which govern the position of primitive elements, which make up the texture [10].

Signal processing methods, such as Wavelet transform [11], [12], [13], Fourier analysis [8] and Gabor filters [14], were motivated by psychophysical researches, which have given evidences that the human brain does a frequency analysis of the image [15], [16]. These approaches represent the texture as an image in a space whose coordinate system has an interpretation that is closely related to the characteristics of a texture. The texture feature coding method (TFCM) that forms the basis for texture features was first discussed by Horng [17] and later applied in various application such as tumor detection and landmine detection [18], [19]. Multi resolution methods such as wavelet transform is widely used and applied in texture analysis. Arivazhagan et al. proposed to use wavelet co-occurrence features and wavelet statistical features to discriminate texture classes [20]. Selvan et al. used singular value decomposition (SVD) on wavelet transform to model a probability density function for texture classification [21]. Moreover, neural networks and machine learning are also applied to learn and classify texture classes by wavelet and information theory features [11], [12], [22]. However, most of texture classification methods are proposed to gray level texture images. Due to the complexity of color space, existing methods cannot perform well classification in color texture images. Therefore, it is an important and imperative work to study on the color texture classification. Linear discriminant analysis is found to be efficiently used in pattern recognition [23], [25].

Authors [24] have proposed NSCT and Local directional binary patterns (LDBP) for feature extraction and k-NN classifier for classification. Shift and rotation invariant texture classification using support vector machine is proposed for texture classification [26]. LDBP captures descriptive textural information. Local directional binary patterns in texture image are obtained by considering different neighborhoods for different radial distances of small kernel. The dominant LDBP's are explored which are fundamental properties of local image texture, while NSCT features aim at supplying additional textural information [27]. The author has proposed a method of a wavelet domain feature extractor and ensembles of neural networks classifier for color texture classification problem [28]. The color texture image classification based on HSV color space, wavelet transform, and motif patterns is implemented and support vector machine (SVM) is applied to learn and classify texture classes [29]. The multiscale techniques for gray level texture analysis is been extended to texture classification of color images by using the multiwavelets transform and Genetic Programming classifier is used [30]. The computer vision community has shown great deal in representing images using multiple scales. The technique is to decompose an image into family of derived images [31], [32]. Perona and Malik [34] formulated the new concept that modified the linear scale space paradigm to smooth within a region while preserving edges. Due to the increasing interest in image analysis, a novel framework to model textures is proposed.

This paper proposes a RGB-based color texture image classification using anisotropic diffusion and LDBP. The anisotropic diffusion is applied to obtain the texture component of image obtained by the difference between original image and cartoon approximation. Further, LDBP approach is used to obtain the features from the texture component. To enhance the class separability, the linear discriminant analysis (LDA) is employed. The classification is performed using k-NN classifier. The experimental results indicate the effectiveness of the proposed method

2   Anisotropic Diffusion

Anisotropic diffusion is a technique where the diffusion is guided by the local structures in the image. It smoothes the image by creating cartoon approximation, which reduces random noise, preserving edges sharp and unmodified. Using anisotropic diffusion, the image is gradually smoothed, while texture component is enhanced by the difference between the original image and cartoon approximation. These filters are difficult to analyze mathematically, as they may act locally like a backward diffusion process. This gives rise to well-posedness questions. On the other hand, nonlinear diffusion filters are frequently applied with very impressive results. Perona and Malik propose a nonlinear diffusion method for avoiding the blurring and localization problems of linear diffusion filtering [34], [35]. They apply an inhomogeneous process that reduces the diffusivity at those locations which have a larger likelihood to be edges. This likelihood is measured by $|\nabla|^2$. The Perona-Malik filter is based on the equation given in Eq. (1):

$$\partial_t u = div \left(g \, |\nabla u|^2 \, \nabla u \right) . \tag{1}$$

and it uses diffusivities as in Eq. (2):

$$g \left(s^2 \right) = \frac{1}{1 + s^2 / \lambda^2} \ (\lambda > 0) . \tag{2}$$

where λ is contrast parameter, which depends on the problem. It utilizes a scalar-valued diffusivity and not a diffusion tensor. There exist a relation between Eq. (1) and neural dynamics of brightness perception.

3   Local Directional Binary Patterns (LDBP)

The most important properties of local directional binary patterns (LDBP) [24] are tolerance against illumination changes and computational simplicity. It extracts local information of texture image. The LDBP transforms signals from spatial representation into frequency representation. An image in LDBP is represented as a sum of sinusoids of changing magnitudes and frequencies. Lower frequencies are more frefrequent than higher frequencies in a particular image. Image is transformed into its frequency component and gives away lot of higher frequency coefficients than the amount of data needed to describe the image. It can be reduced without compromising on image quality. Most of the visually significant information about the image is only concentrated in just a few coefficients and LDBP makes use of this property.

The basic idea of LDBP is that, 3x3 kernel of image can be treated as basic texture region. The gray value of central pixel is compared with the gray values of eight pixels around it. The central gray pixel value is the threshold value. If the gray value of surrounding pixel is larger than gray value of central pixel, the surrounding pixel is marked as one otherwise zero. The binary values of all surrounding pixels can be obtained. All surrounding pixels are given different metrics. The metrics is multiplied with a binary value of surrounding pixels as shown in the Fig. 1. Further, the sum of product of binary value and metrics of all surrounding pixels is set as the value of local directional binary pattern of central pixel. The value of local directional binary patterns of all pixels in image can be obtained through such calculation neglecting the pixels of edges. The LDBP weight can be calculated using the Eq. (3):

$$f_b \left(x_c, x_c \right) = \sum_{j=0}^{7} v \left(f_j - f_c \right) \cos \left(j * 45 \right) . \tag{3}$$

where f_c is gray value of the central pixel of local texture region, f_j is the gray value of jth surrounding pixel of local texture region.

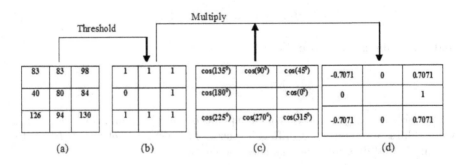

Fig. 1. Transformation of neighborhood pixels to calculate central pixel weight in LDBP. (a) A sample neighborhood, (b) Resulting binary thresholded result, (c) LDBP mask, (d) Resultant weights after multiplying corresponding elements of (b) and (c).

4   The Proposed Method for Texture Image Classification

The proposed method for texture image classification consists of two modules, namely, texture training module and classification module.

4.1   Texture Training Module

Each block of image is decomposed with anisotropic diffusion [34] to obtain texture approximation. The LDBP weights of each texture approximation are calculated. The Haralick features, namely, contrast, energy, entropy, homogeneity, maximum probability, cluster shade and cluster prominence are obtained from LDBP matrix. The steps of the proposed method are given in the Algorithm 1.

Algorithm 1:   Training Algorithm
Step 1 : Input the training image block I of size 64 x 64.
Step 2 : Extract the red component of the image I.
Step 3 : Decompose the image I with anisotropic diffusion.
Step 4 : Obtain texture approximation I_{txr}.
Step 5 : Compute LDBP weights for image I_{txr} in Step 4, compute Haralick features (7 features x 4 angles = 28 numbers), to obtain feature vector F and store F in the feature database.
Step 6 : Repeat the Steps 1-5 for green, blue components of all the training blocks of all the texture class images and Obtain the training set (TF) of feature vectors.
Step 7 : Apply LDA on training feature set (TF) of Step 6 to obtain discriminant feature set (TFLDA). Store TFLDA in the feature library, which is to be used for texture classification. Denote TFLDA vector as $f_{(lib)}$ vector.
Step 8 : Stop.

4.2   Texture Classification

If $f_{(test)}(x)$ is the feature vector of test sample image x and $f_{(lib)}(m)$ is the feature vector of the mth class in the feature library, then the Euclidean distance between these two vectors is given by the Eq. (4) :

$$D\left(f_{test}, f_{lib}\right) = \sqrt{\sum_{i=1}^{N}\left(f_{(test)i}(x) - f_{(lib)i}(m)\right)^2} \qquad (4)$$

where N is the number of features in the feature vector. The image x is classified using the k nearest neighbor (k-NN) classifier [35]. In k-NN classifier, the class of the test sample is decided by the majority class among the k nearest neighbors. A neighbor is decided to be nearest if it has the smallest distance in the feature space. In order to avoid a tied vote, it is preferable to choose k to be an odd number. The experiments are carried out using k-NN classifier with k=3. The testing algorithm is given in the Algorithm 2.

Algorithm 2 : Testing Algorithm (Classification of test images)

Step 1 : Input the testing image block I_{test} of size 64 x 64.

Step 2 : Extract the red component of the image I_{test}.

Step 3 : Decompose the image I_{test} with anisotropic diffusion.

Step 4 : Obtain texture approximation $I_{testtxr}$.

Step 5 : Compute LDBP weights for image $I_{testtxr}$ in Step 4, to compute Haralick features (7 features x 4 angles = 28 numbers).

Step 6 : Repeat the Steps 1 - 5 for green, blue component of the image to obtain feature vector F_{test} and store F_{test} in the feature database.

Step 7 : Project F_{test} on TFLDA components and obtain the weights $F_{testLDA}$ which are considered as reduced test image features. Denote $F_{testLDA}$ as $f_{(test)}$.

Step 8 : (Classification) Apply k-NN classifier (k = 3) to classify the test image I_{test} as belonging to class m (m = 1,2,..,16), using Euclidean distance.

Step 9 : Stop.

5   Experimental Results and Discussion

The experimentation is carried out using sixteen texture classes (Fig. 2) [30] of Oulu texture album [36], each with an image of size 512 x 512. Each texture image is dividing into 64 equal sized non overlapping blocks of size 64 x 64, out of which 32 randomly chosen blocks are used as the training samples and remaining 32 blocks are considered as test samples for each texture class. Thus, a total of 1024 texture images are considered. The ten fold experiments are carried out by randomly selecting training and testing samples and the average classification results are obtained.

The experimentation of the proposed method is carried out on Intel(R) Core(TM) i3-2330M @ 2.20GHz with 4 GB RAM using MATLAB 7.9 software. We apply anisotropic diffusion [34], which is used to smooth image while preserving edges. The original image (f) is decomposed in a set of derived images with anisotropic diffusion. The procedure is executed with several levels of decomposition (t), wherein we obtain two components: cartoon approximation (u) and texture approximation (v). The texture component is obtained by subtracting the original image and the cartoon approximation. The filtering process aims at evidencing high frequencies in the image in order to produce richer representation, which overcomes the main restriction imposed by linear approaches, i.e. blur in region boundaries. The texture image v is then used to compute LDBP weights.

The LDBP approach is used to obtain the rotation invariant coefficients of the image (of size 62 x 62). The operator labels the pixels of an image by thresholding a 3 x 3 neighborhood of each pixel with center value and yields the results as binary numbers. Further, this matrix is used to obtain LDBP weights, in order to compute Haralick features over the angles 0^0, 45^0, 90^0 and 135^0 for each color channel, which leads to 84 (28 features x 3 channels) features. The class separability is enhanced using LDA, which reduces the

Fig. 2. Color texture images from left to right and top to bottom: Grass, Flowers1, Flowers2, Bark1, Clouds, Fabric7, Leaves, Metal, Misc, Tile, Bark2, Fabric2, Fabric3, Food1, Water, and Food2

dimension of the feature vector to 15. The k-NN classifier with k=3, is used for classification. An exhaustive search for best parameters like levels of decomposition and lambda (λ) is investigated. Experiments are carried out for different levels of decomposition ranging from 10 to 200 and values of λ ranging from 0.05 to 0.25. The Table 1 shows the average classification accuracy, training time and testing time for 20 levels of decomposition for varying values of lambda (λ). It plays the role of a contrast parameter separating forward (low contrast) from backward (high contrast) diffusion areas. It is observed from the Table 1 that the average classification accuracy for λ initially increases, reaches maximum value of 99.80% with training time of 497.6807 sec. and testing time of 26.1151 sec. for λ=0.10 and then decreases. Hence, optimal value λ =0.10 is used for further experimentation.

Table 1. Average Classification Accuracy (%), Training Time and Testing Time for Varying Values of lambda (λ) for 20 Levels of Decomposition

λ	Average Classification Accuracy (%)	Training Time (sec.)	Testing Time (sec.)
0.05	99.414063	498.7848	26.5662
0.10	99.804688	497.6807	26.1151
0.15	99.023438	497.1358	26.1499
0.20	99.414063	498.6073	27.1175
0.25	99.218750	498.7649	27.51376

The Table 2 shows the average classification accuracy (%), training time and testing time of proposed method for different levels of decomposition in anisotropic diffusion for λ=0.10. It is observed from Table 2 that the average classification accuracy increases to a maximum value of 99.804688% with training time of 497.6807 sec. and testing time of 26.1151 sec. for 20 levels of decomposition and then remain same for further levels of decomposition. Hence, optimized value 20 for levels of decomposition is considered for experimentation. The Table 3 shows the classification accuracies for each texture class after applying the optimized value of λ=0.10 and levels of decomposition = 20. Table 3 shows comparison of results obtained by the proposed method and other methods in the literature [30], [29], [21], [11]. Methods of [30], [29], [21], [11] are experimented on the same dataset [36] as is used in the proposed method. So that, results can be compared. It is noticed

Table 2. Average Classification Accuracy (%), Training Time and Testing Time of Proposed Method for Different Levels of Decomposition in Anisotropic Diffusion (λ=0.10)

Levels of Decomposition	Average Classification Accuracy (%)	Training Time (sec.)	Testing Time (sec.)
10	99.609375	498.0112	26.2111
20	99.804688	497.6807	26.1151

Table 3. Comparisons of Classification Rates (%) of the Proposed Method and Other Methods in the Literature

Sl. No.	Image Name	The Proposed Method	Abdulmunim [30]	Chang et al.[29]	Sengur et al.[11]	Selvan and Ramakrishnan [21]
1	Grass	100	100	96	92	86
2	Flowers1	96.875	100	99	98	90
3	Flowers2	100	100	96	90	89
4	Bark1	100	97	98	94	90
5	Clouds	100	98	100	100	97
6	Fabric7	100	100	97	91	92
7	Leaves	100	100	100	98	92
8	Metal	100	100	93	90	90
9	Misc	100	100	99	94	93
10	Tile	100	99	97	94	95
11	Bark2	100	100	97	92	90
12	Fabric2	100	95	100	99	97
13	Fabric3	100	98	100	99	99
14	Food1	100	95	99	97	95
15	Water	100	99	100	100	100
16	Food2	100	100	95	89	90
Mean success rate (%)		99.8047	98.8125	97.875	94.8125	92.8125

from the Table 3 that the proposed method offers better classification performance. The proposed method with anisotropic diffusion and LDBP yields 99.8047% mean success rate. Further, in general, the SVM classifier gives better results than k-NN classifier. However, the textural features extracted in the proposed method are so richer that even a simple k-NN classifier yields better results than other feature set with SVM classifier. The SVM classifier used in Chang et al.[29] has given poorer results (97.87%) than the proposed method with k-NN classifier (99.80%). The classification results of the proposed method can still be improved by using SVM classifier.

The experimental results indicate that the proposed method is effective in terms of classification accuracy and reduced feature set.

6   Conclusion

In this paper, RGB based color texture image classification using anisotropic diffusion and LDBP is proposed. The richness of texture attribute is enhanced by anisotropic diffusion. Unlike other methods in literature, the proposed technique processes the texture data directly in order to preserve useful information in the image. The LDBP co-occurrence features represent the texture characteristics. The class discriminality is enhanced using LDA. The features obtained from LDA are representative of each class. The classification performance is tested on Oulu textures. The k-NN classifier with k=3 is used to classify images. Experimental results demonstrate the effectiveness of proposed method. From experimental results, it is observed that the proposed method outperforms other methods in the literature with computational simplicity and reduced features for texture classification.

Acknowledgments. Authors are indebted to the reviewers for their valuable comments and suggestions, which substantially improved the quality of the paper.

References

1. Tuceryan, M., Jain, A.K.: Texture Analysis. In: Handbook of Pattern Recognition and Computer Vision, pp. 235–276 (1993)
2. Haralick, R.M.: Statistical and Structural Approaches to Texture. Proceedings of IEEE 67(5), 786–804 (1979)
3. Zhang, J., Tan, T.: Brief Review of Invariant Texture Analysis Methods. Pattern Recognition 35(3), 735–747 (2002)
4. Petrou, M., García-Sevilla, P.: Image Processing Dealing with Texture. Wiley (2006)
5. Mirmehdi, M., Xie, X.H., Suri, J.: Handbook of Texture Analysis. World Scientific (2008)
6. Chen, Y.Q., Nixon, M.S., Thomas, D.W.: Statistical Geometrical Features for Texture Classification. Pattern Recognition 28(4), 537–552 (1995)

7. Vilnrotter, F.M., Nevatia, R., Price, K.E.: Structural Analysis of Natural Textures. IEEE Transactions on Pattern Analysis and Machine Intelligence 8(1), 76–89 (1986)
8. Azencott, R., Wang, J.-P., Younes, L.: Texture Classification using Windowed Fourier Filters. IEEE Transactions on Pattern Analysis and Machine Intelligence 19(2), 148–153 (1997)
9. Julesz, B.: Experiments in the Visual Perception of Texture. Scientific American 232(4), 34–43 (1975)
10. Keller, J.M., Chen, S., Crownover, R.M.: Texture Description and Segmentation Through Fractal Geometry. Computer Vision Graphics and Image Processing 45(2), 150–166 (1989)
11. Sengur, A., Turkoglu, I., Ince, M.C.: Wavelet Packet Neural Networks for Texture Classification. Expert Systems with Applications 32(2), 527–533 (2007)
12. Sengur, A.: Wavelet Transform and Adaptive Neuro-fuzzy Inference System for Color Texture Classification. Expert Systems with Applications 34(3), 2120–2128 (2008)
13. Karabatak, M., Ince, M.C., Sengur, A.: Wavelet Domain Association Rules for Efficient Texture Classification. Applied Soft Computing 11(1), 32–38 (2011)
14. Daugman, J., Downing, C.: Gabor Wavelets for Statistical Pattern Recognition. In: Arbib, M.A. (ed.) The Handbook of Brain Theory and Neural Networks, pp. 414–419. MIT Press, Cambridge (1995)
15. Daugman, J.G.: Uncertainty Relation for Resolution in Space, Spatial Frequency and Orientation Optimized by Two-Dimensional Visual Cortical Filters. Journal of the Optical Society of America 2(7), 1160–1169 (1985)
16. Mayhew, J.E.W., Frisby, J.P.: Texture Discrimination and Fourier Analysis in Human Vision. Nature 275, 438–439 (1978)
17. Horng, M.H.: Texture Feature Coding Method for Texture Classification. Opt. Eng. 42(1), 228–238 (2003)
18. Liang, J., Zhao, X., Xu, R., Kwan, C., Chang, C.-I.: Target Detection with Texture Feature Coding Method and Support Vector Machines. In: Proc. ICASSP, Montreal, QC, Canada, pp. II-713–II-716 (2004)
19. Torrione, P., Collins, L.M.: Texture Features for Antitank Landmine Detection using Ground Penetrating Radar. IEEE Trans. Geosci. Remote Sens. 45(7), 2374–2382 (2007)
20. Arivazhagan, S., Ganesan, L.: Texture Classification using Wavelet Transform. Pattern Recognition Letters 24(9-10), 1513–1521 (2003)
21. Selvan, S., Ramakrishnan, S.: SVD-based Modeling for Image Texture Classification using Wavelet Transformation. IEEE Transactions on Image Processing 16(11), 2688–2696 (2007)
22. Turkoglu, I., Avci, E.: Comparison of Wavelet-SVM and Wavelet-Adaptive Network based Fuzzy Inference System for Texture Classification. Digital Signal Processing 18(1), 15–24 (2008)
23. Ghafoor, A.: Multimedia Database Management System. ACM Comput. Surv. 27(4), 593–598 (1995)
24. Hiremath, P.S., Bhusnurmath, R.A.: Texture Image Classification using Nonsubsampled Contourlet Transform and Local Directional Binary Patterns. Int. Journal of Applied Research in Computer Science and Software Engineering 3(7), 819–827 (2013)
25. Shivashankar, S., Hiremath, P.S.: PCA plus LDA on Wavelet Co-Occurrence Histogram Features for Texture Classification. Int. Journal of Machine Intelligence 3(4), 302–306 (2011)

26. Hiremath, P.S., Bhusnurmath, R.A.: Nonsubsampled Contourlet Transform and Local Directional Binary Patterns for Texture Image Classification Using Support Vector Machine. Int. Journal of Engineering Research and Technology 2(10), 3881–3890 (2013)

27. Hiremath, P.S., Bhusnurmath, R.A.: A Novel Approach to Texture Classification using NSCT and LDBP. IJCA Special Issue on Recent Advances in Information Technology (NCRAIT 2014) 3, 36–42 (2014) ISBN-973-93-80880-08-3

28. Sengur, A.: Color Texture Classification using Wavelet Transform and Neural Network Ensembles. The Arabian Journal for Science and Engineering 34(2B), 491–502 (2009)

29. Chang, J.-D., Yu, S.-S., Chen, H.-H., Tsai, C.-S.: HSV-based Color Texture Image Classification using Wavelet Transform and Motif Patterns. Journal of Computers 20(4), 63–69 (2010)

30. Abdulmunim Matheel, E.: Color Texture Classification using Adaptive Discrete Multiwavelets Transform. Eng. & Tech. Journal 30(4), 615–627 (2012)

31. Lindeberg, T.: Scale-space. In: Wah, B. (ed.) Encyclopedia of Computer Science and Engineering, EncycloCSE 2008, vol. 4, pp. 2495–2504. John Wiley and Sons, Hoboken (2008)

32. Witkin, P.: Scale-Space Filtering. In: Int. Joint Conference on Artificial Intelligence, pp. 1019–1022 (1983)

33. Perona, P., Malik, J.: Scale Space and Edge Detection using Anisotropic Diffusion. In: Proc. IEEE Comp. Soc. Workshop on Computer Vision, Miami Beach, November 30-December 2, pp. 16–22. IEEE Computer Society Press, Washington (1987)

34. Perona, P., Malik, J.: Scale Space and Edge Detection using Anisotropic Diffusion. IEEE Trans. Pattern Anal. Mach. Intell. 12, 629–639 (1990)

35. Duda, R.O., Hart, P.E.: Stork: Pattern Classification. Wiley Publications, New York (2001)

36. Internet: University of Oulu texture database (2005), http://www.outex.oulu.fi/outex.php

A Knowledge-Based Design for Structural Analysis of Printed Mathematical Expressions

Pavan Kumar P.*, Arun Agarwal, and Chakravarthy Bhagvati

School of Computer and Information Sciences
University of Hyderabad, Hyderabad 500 046, India
pavan.ppkumar@gmail.com, {aruncs,chakcs}@uohyd.ernet.in

Abstract. Recognition of Mathematical Expressions (MEs) is a challenging Artificial Intelligence problem as MEs have a complex two dimensional structure. ME recognition involves two stages: Symbol recognition and Structural Analysis. Symbols are recognized in the first stage and spatial relationships like superscript, subscript etc., are determined in the second stage. In this paper, we have focused on structural analysis of printed MEs. For structural analysis, we have proposed a novel ternary tree based representation that captures spatial relationships among the symbols in a given ME. Proposed tree structure has been used for validation of generated ME structure. Structure validation process detects errors based on domain knowledge (mathematics) and the error feedback is used to correct the structure. Therefore, our validation process incorporates an intelligent mechanism to automatically detect and correct the errors. Proposed approach has been tested on an image database of 829 MEs collected from various mathematical documents and experimental results are reported on them.

Keywords: Mathematical expressions, structural analysis, ternary tree representation, domain knowledge, structure validation.

1 Introduction

Mathematical Expressions (MEs) form a significant part in scientific and engineering disciplines. MEs can be offline or online. Offline or Printed MEs take the form of scanned images while online MEs are written using data tablets. ME recognition is the process of converting printed or online MEs into some editable format like LaTeX, MathML etc. It is needed for applications like digitizing scientific documents, generating braille script for visually impaired etc. ME recognition involves two stages: Symbol recognition and Structural analysis. Symbols are recognized in first stage and structure (spatial relationships) is interpreted in second stage. As mentioned in [3], structural analysis plays a vital role in the overall ME recognition task.

In [11], we have discussed that mathematical symbols can be composed of one or more indivisible units called *Connected Components (CCs)*. For example,

---

* Corresponding author.

M.N. Murty et al. (Eds.): MIWAI 2014, LNAI 8875, pp. 112–123, 2014.

= is composed of two horizontal line CCs. For some symbols, identities depend on context. For example, a horizontal line CC can be MINUS, OVERBAR, FRACTION etc., as its identity depends on neighbouring symbols (context). If it has symbols above and below, it is a FRACTION. If it has symbols only below, it is an OVERBAR and so on. Symbols that are composed of more than one CC are called *Multi-CC* symbols. Symbols whose identities depend on context are called *Context-dependent* symbols. In [11], we have shown an architecture for structural analysis of printed MEs. This design comprises three modules: Symbol formation, Structure generation and Generation of encoding form like LATEX. Symbol formation process takes labelled CCs of an ME image as input and forms Multi-CC symbols as well as resolves the identity of Context-dependent symbols. Elements of Multi-Line MEs (MLMEs) like matrices and enumerated functions are also extracted in this module. Structure generation module takes the formed symbols, analyzes spatial relationships like superscript, subscript etc., among them and generates an intermediate tree representation for the given ME. Third module generates an encoding form like LATEX by traversing ME tree.

In [11], we have discussed only symbol formation process. In this paper, we have discussed the other two modules. We have also added a new *structure validation* module that automatically detects and corrects the errors before encoding form is generated. The paper is organized as follows: Existing works on structural analysis are discussed in Section 2. Section 3 gives an overall design of our approach to structural analysis. Structure generation module is presented in Section 4. Section 5 presents structure validation process along with encoding form generation. Experimental results are summarized in Section 6 and the paper is concluded in Section 7.

2   Related Work

In [18], a recent survey on ME structural analysis can be found. Existing works on structural analysis along with their intermediate representations are discussed below.

2.1   Intermediate Representation

Lee et al. [8] have proposed a data structure that captures spatial relations among the symbols of an ME. Each symbol is represented by a structure that has label information which gives identity of the symbol, and six pointers. These six pointers are meant to point to six spatially related symbols. Each symbol forms a node in the tree which is formed by joining pointers of the symbols appropriately. In this structure, most of the pointers are empty and the tree is sparse. Hence the data structure is not spatially efficient as well as it cannot handle all types of MEs like MLMEs. In [14], an ME has been represented in the form of a directed graph. Each node of the graph corresponds to a symbol and a link between two nodes has the following information: Labels of $node_1$ and $node_2$ along with spatial relationship between them (one of the above mentioned regions). As only links are stored, number of links to be used depends on number of possible relationships among the symbols of an ME, and hence not fixed and also not known a priori. In addition, it also cannot handle MLMEs. Zanibbi et al. [19] have proposed a tree structure that contains two types of nodes: Symbol nodes and Region nodes. Symbol node represents a symbol and stores its identity. Region node is a child of symbol node that represents symbols that are spatially related (excluding the horizontally adjacent relation) to symbol node and the type of spatial relation (ABOVE, BELOW, SUPERSCRIPT, SUBSCRIPT etc.) is stored in the node. The children of region node are again symbol nodes whose symbols are represented by it and all these symbols are horizontally adjacent. Tree has region and symbol nodes at alternate levels. To handle MLMEs, the authors have extended their tree structure [17] to have TA-BLE (to designate MLMEs), ROW (for rows) and ELEM (for elements) nodes. In this tree structure, number of children for region nodes is not known a priori as it depends on the number of symbols in that subexpression.

2.2   Structure Generation

Structure generation process analyzes spatial relations among the symbols of a given ME and generates its intermediate representation. In [6], LaTeX code is directly generated after analyzing spatial relations without using any intermediate representation. Lee et al. [8] have proposed a method in which special mathematical symbols like $\sum$, $\int$ etc., are analyzed first, then matrices are detected and finally superscripts and subscripts of remaining symbols are captured. They have detected and extracted the elements of matrices as part of structure generation. But their approach has not been discussed in a detailed manner. Tian et al. [16] have performed structure generation of offline MEs using baseline information. In [5], a network that represents spatial relations has been constructed and minimal spanning tree that corresponds to the actual structure has been obtained. Zanibbi et al. [19] have proposed to construct baseline structure tree for an ME based on reading order and operator dominance. [15] have presented a method based on a minimum spanning tree construction and symbol dominance for online handwritten MEs. Several approaches [1,6,9] have used grammars to generate structures and hence difficult to tackle erroneous ones.

3   Proposed Design to ME Structural Analysis

Proposed architecture to ME structural analysis is shown in Fig. 1. For a given input ME image, it is binarized (converted from gray scale to binary image using Otsu's method [7]), CCs are extracted [7] and labels are assigned to them. In the above process, Minimum Bounding Rectangles (MBRs) of CCs are also computed. MBR of a CC is the minimum rectangle bounding it and is represented by its top-left and bottom-right co-ordinates. As shown in Fig. 1, proposed approach starts with labelled CCs and comprises four modules namely symbol formation, structure generation also called as *ME tree* generation, structure validation and generation of encoding form like LaTeX. Structure validation module

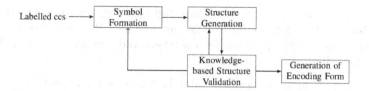

Fig. 1. Proposed architecture for ME Structural Analysis

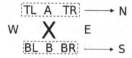

Fig. 2. Spatial regions around a mathematical symbol **X**. TL, A, TR, BL, B and BR denote top-left, above, top-right, bottom-left, below and bottom-right regions respectively. N – Northern region, S – Southern region, E – East, W – West.

inspects tree structure and gives error feedback (shown as arrows) using domain knowledge, to symbol formation and structure generation modules. Symbol formation and structure generation processes are repeated until no errors are found by validation module, after which encoding form is generated.

As discussed in [11], symbol formation process takes labelled CCs as input and forms Multi-CC and Context-dependent symbols. MLMEs are also detected and their elements are extracted. To handle MLMEs, starting and ending delimiters like $($, $[$, $\{$, $|$ and $)$, $]$, $\}$, $|$ respectively are considered as Context-dependent symbols as they can be used to enclose sub-expressions or MLMEs. Based on horizontal and vertical projection profiles [11], CCs between the delimiters are analyzed to compute rows and elements. If there is only one row and element, it is not an MLME. Otherwise, labels of the delimiters are changed to those of MLME ones and all the computed elements in each row are isolated (but their association with the delimiters is stored) from the main stream of CCs. For example, $\div$ is a Multi-CC symbol and has three CCs (one horizontal line and two DOTs). If these three CCs are vertically stacked one over another, their MBRs are combined to form a single composite symbol. Label corresponding to $\div$ is assigned to the composite symbol.

4   ME Tree Structure and Its Generation

Structure generation module takes left to right ordered symbols (from symbol formation process), analyzes spatial relations and generates an *ME tree* proposed for the purpose. In [13], we have discussed that symbols in MEs can have surrounding symbols in its top-left, above and top-right regions as well as in their bottom-left, below and bottom-right regions. For example, in $\sum_i^j$, symbols i and j are in bottom-right and top-right of $\sum$ respectively. The region formed by combining top-left, above and top-right regions (bottom-left, below and bottom-right) in that order is called as *Northern (Southern) region*. These regions are shown in Fig. 2. Symbols in northern and southern regions of X align in its vertical direction.

Definition 1. A mathematical operator is called as *Horizontal (H) operator* if it does not have symbols in their northern and southern regions (Eg: $+$, $-$, $<, \leq, \geq, =$ etc.). Otherwise, it is a *Non-Horizontal (NH) operator* (Eg: $\sum, \int$, FRACTION etc., and accent symbols like HAT, OVERBAR etc.).

Definition 2. A symbol which is not present either in the northern or in the southern region of any other symbol in an ME is called *Baseline symbol*. For example, in $a^2 + b^2 + 2ab$, baseline symbols are: a, $+$, b, $+$, 2, a, b. Remaining two symbols 2 and 2 are in the top-right regions of a and b respectively.

4.1   Tree Representation

Proposed representation uses a ternary tree structure. Each symbol is represented using only three pointers to represent spatial relationships around it. Out of the three, one pointer is meant to point to the next baseline symbol and the other two are meant for the entire northern and southern regions respectively. In most of the MEs, the entire northern or southern region for any symbol forms a single subexpression. There may be unusual cases, where two different subexpressions are present in the same region (Eg: $_nC_r$). There can also be rare cases where symbols can have more than two different subexpressions (if they have pre-super and pre-sub scripts in addition to super and sub-scripts) over the northern and southern regions, but occur rarely in mathematics. Our representation handles unusual and rare cases in a different manner (discussed later).

Proposed tree node structure to represent a symbol is given by an abstract data type called as *TreeNode* with some fields. Each symbol in a given ME forms a node in the tree that is generated by linking pointers of the symbols based on their spatial relations. Each field in the data structure is discussed below:

1. Integer field, *label* gives the identity of a symbol.
2. Two boolean fields, *EOE (End of Element)* and *EOR (End of Row)* are used to handle MLMEs like matrices, enumerated functions etc. *EOE* is set to TRUE if a symbol designates end of some element of an MLME. Otherwise, it is set to FALSE. Similarly, *EOR* is set to TRUE if a symbol designates end of some row of an MLME. Otherwise, it is set to FALSE.
3. TreeNode pointer *next* of any symbol points to its next baseline symbol.
4. *nLink* of a symbol points to first baseline symbol of northern expression.
5. *sLink* of a symbol points to first baseline symbol of southern expression.
6. Northern and Southern regions for different symbols are listed below:
   (a) FRACTION – Numerator (denominator) corresponds to northern (southern) region.
   (b) SQUAREROOT – *Degree* (contained expression) corresponds to northern (southern) region.

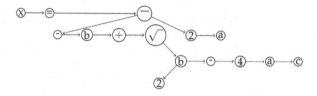

Fig. 3. Proposed tree structure for $x = \frac{-b+\sqrt{b^2-4ac}}{2a}$. Here, x is head of the tree

(c) Accent or Wide accent symbols – Enclosed subexpression is present below (southern) for accent symbols like OVERBAR, OVERBRACE etc., and above (northern) for accent symbols like UNDERBAR, UNDER-BRACE etc.

(d) Other symbols – northern (southern) region gives superscript (subscript).

In our representation, *label* information of a symbol is exploited to resolve its northern and southern regions. First node (root) of the tree (first baseline symbol) is called *head* of the tree. That means, *nLink* and *sLink* pointers point to heads of the subtrees for the northern and southern subexpressions respectively. As northern and southern regions are considered as a whole, *proposed tree structure is simple*. Proposed tree structure for an example ME is shown in Fig. 3. In this figure (and in the subsequent figures), *nLink* is shown by left link, *sLink* by right link and *next* by horizontal link.

4.2 Logical Proof for Completeness

Proposed tree structure handles unusual and rare cases in the following manner.

1. If two different sub-expressions are present in the same (northern or southern) region, one of them is logically shifted to the other region. Logically shifted regions are represented by negating the label of first baseline symbol of the corresponding region. For example, in $_nC_r$, if n is logically shifted, its label is negated and considered as northern child (pointed by *nLink*) of C.

2. Proposed ME representation handles rare cases using special nodes called *ϵ-nodes*. ϵ-nodes are used to handle symbols with more than two different subexpressions over their northern and southern regions. An ϵ-node also has a unique *label*, but does not refer to any symbol. Its *nLink* and *sLink* pointers can point to two more sub-expressions of a symbol, if the symbol has more than two sub-expressions around it. If a rare symbol has more than two and less than or equal to four different subexpressions around it, one ϵ-node is needed. If it has more than four different subexpressions, two ϵ-nodes are needed. ϵ-nodes are connected to the actual symbol node using *next* pointers. For example, let us consider a symbol with four sub-expressions $_q^p\sum_r^s$ around it. Here, super and sub scripts r and s, are pointed by $\sum$ and its pre-super and pre-sub scripts (top-left and bottom-left) p and q, are pointed by an ϵ-node. The ϵ-node is connected to $\sum$ using its *next* pointer which is shown

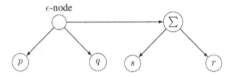

Fig. 4. Proposed tree for $_q^p \sum_r^s$ where $\sum$ has pre-super and sub scripts

in Fig. 4. Let us consider another rare complex example: $_b^a \prod_{i=1}^{n} {}_c^d$. Here, $\prod$ has six different subexpressions around it. In these cases, two ϵ-nodes are used, one to hold top-left and bottom-left subexpressions and the other one to hold above and below subexpressions. Symbol node handles the remaining two subexpressions and ϵ-nodes are connected using *next* pointers. These rare cases almost do not occur in any ME but LaTeX can generate them.

3. **Representing MLMEs:** Our proposed representation handles MLMEs by generating trees for all the elements (as each element is again an ME) recursively in all the rows and attaching them in a row-major order. That means, *next* of starting delimiter (like (, [etc.) of an MLME points to head of tree for first element in the first row. For any element in any row, *next* of its last baseline symbol points to head of tree for the next element in that row and *EOE* is set to TRUE for this last baseline symbol. For last element in any row, *next* of its last baseline symbol points to head of tree for first element of next row and *EOR* is set to TRUE for this last baseline symbol. Pointer *next* of last baseline symbol of last element in last row points to the ending delimiter (like),] etc.) if present (not present for enumerated functions). It is to be noted that the above *recursive process handles even nested MLMEs*.

4.3   Spatial Efficiency and Generality

As northern and southern subexpressions are taken as a whole, processing complexity of the proposed tree is reduced. To handle MLMEs, only two bits are used. Therefore, proposed tree is spatially efficient. In general, mathematical symbols have atmost two arguments (numerator/denominator for fraction, degree/contained expression for squareroot, super/subscripts for others etc.) and so our approach is intuitive. Proposed tree structure can also be used to handle other structures that are similar in nature to MEs. For example, chemical equations have ionic information (like oxidation state) and the number of instances of an atom in the northern and southern regions. In [12], LaTeX based linear representation has been used for ME retrieval. As proposed tree structure is simple and complete, its linear form (symbol, its northern and southern expressions in that order recursively) gives a better representation for this application.

4.4   Algorithm to Generate ME Tree

Proposed algorithm takes left to right ordered symbols of an ME as input, generates ME tree and returns a pointer to *head* of the tree:

TreeNode * generateTree(L: list of Symbols)

1. Find the baseline symbols in L by isolating their northern and southern subexpressions using the next two steps. These subexpressions are first isolated for NH operators and then for other symbols (as northern and southern symbols of NH operators are present on their both sides).
2. Scan list L from left to right and if a NH operator say X, is found, do:
   (a) Create a tree node of type *TreeNode* for X.
   (b) Inspect the symbols on both sides of X in L and isolate symbols in its northern and southern regions. If a line segment obtained by joining the midpoints or centroids of MBRs of two symbols X and Y, is almost vertical (discussed earlier), then Y is in the northern or southern region of X. In our implementation, angles from $45°$ to $90°$ are considered almost vertical.
   (c) Isolated vertically aligned symbols are partitioned into northern and southern ones based on their position with respect to X. If an isolated symbol is present in the above (below) of X, then it is in the northern (southern) region of X.
   (d) If more than two subexpressions are present over the northern and southern regions, ϵ-nodes are accordingly added and those subexpressions are attached to them.
3. Scan the remaining symbols (other than NH operators and their corresponding isolated symbols in step 2) in L from left to right and for each such symbol, say X:
   (a) Create a node of type *TreeNode* for X.
   (b) Isolate symbols (using vertical alignment criteria) in its northern and southern regions by inspecting on the right side of the current symbol.
4. Connect the nodes created in steps (2) and (3) (baseline symbols) using *next* pointers. Let pointer to *head* of the tree be denoted by H.
5. Traverse through the baseline symbol nodes and for each such node, generate trees for their northern and southern subexpressions recursively:
   (a) Let pointer to the current baseline symbol node be denoted by X (Initially, $X = H$). Generate ME tree for northern region of X, if present, and assign pointer to its head to $nLink$ of X. Let symbol list in the northern region (captured either in step (2) or (3) above) be denoted by NR. Therefore, $X \rightarrow nLink = generateTree(NR)$.
   (b) Similarly, $X \rightarrow sLink = generateTree(SR)$, where SR denotes symbol list in the captured southern region.
   (c) Go to the next baseline symbol node. $X = X \rightarrow next$. Go to step 5(a).
6. Traverse through baseline symbol nodes and if any MLME delimiter is found, generate and attach trees of its elements in a row-major order recursively.
7. Return *head* of the final tree generated for the given ME. *return H.*

$$a(i,j) = \begin{cases} \boxed{2^j} & \boxed{i=1} \\ \boxed{a(i-1,2)} & \boxed{j=1} \\ \boxed{a(i-1,a(i,j-1))} & \boxed{i,j \geq 2} \end{cases}$$

Fig. 5. An enumerated function image with its multi-CC symbols and extracted elements (shown in boxes) after symbol formation process

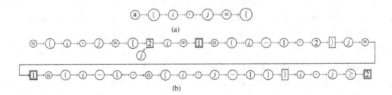

(a)

(b)

Fig. 6. ME tree generation for the enumerated function shown in Fig. 5 (a) Initial tree (b) Final tree after the elements are attached

An enumerated function image is shown in Fig. 5, in which, Multi-CC symbols are formed as well as its elements are extracted in the symbol formation process (shown in boxes). Its ME tree generation process is shown in Fig. 6, in which intial and final trees (before and after element attachment) are shown in Fig. 6(a) and (b) respectively. For rectangular nodes, EOE is set to TRUE and for double rectangular nodes, EOR is set to TRUE.

5   Structure Validation

Generated ME tree is validated using domain (mathematics) knowledge to verify correctness of the structure. If any erroneous structure is detected, corresponding feedback is given to symbol formation or structure generation modules, according to the source of error. Common errors that occur in the symbol formation module are Multi-CC as well as Context-dependent symbols may not be correctly resolved. Similarly in structure generation module, super or sub-script relationships may be lost for a symbol. Our validation algorithm is given below:

1. ME tree is traversed in pre-order [4] (visit in the following order: current node, its $nLink$, $sLink$ and $next$ recursively) and verified for errors using domain knowledge. If no errors are found, encoding form is directly generated. Otherwise, validation module decides if the errors are from symbol formation or structure generation modules.
2. If an error occurs in symbol formation module, feedback is given to that module and that process is repeated again by taking that error into account. Newly formed symbols are sent to structure generation module and the new tree structure generated is again validated.
3. If the error occurs in structure generation module, a new tree structure is generated by taking into account the error feedback.
4. The above two steps are repeated until no errors are found.

$$f'(a) = \frac{df}{dx}\bigg|_{x=a} = \frac{dy}{dx}\bigg|_{x=a} = Df(a)$$

(a)

```
f'(a)= \frac {df}{dx}
\left|\begin{array}{cc}
dy \\ - \\ - \\ - \\
x=a & dx \\ \end{array}
\right|_{x=a}=Df(a)
```

$$f'(a) = \frac{df}{dx}\left|\begin{array}{c} dy \\ - \\ - \\ - \\ x = a \quad dx \end{array}\right|_{x=a} = Df(a)$$

(b)   (c)

```
f'(a)= \frac {df}{dx}
|_{x=a}= \frac {dy}{dx}
|_{x=a}=Df(a)
```

$$f'(a) = \frac{df}{dx}\big|_{x=a} = \frac{dy}{dx}\big|_{x=a} = Df(a)$$

(d)   (e)

Fig. 7. (a) An ME image (b) LATEX output without validation (c) Regenerated ME from LATEX output (b) (incorrect) (d) LATEX output with error feedback to symbol formation process (e) Regenerated ME from LATEX output (d) (corrected)

Validation module uses domain knowledge (mathematical properties encoded in the form of rules) to detect errors in the tree. The knowledge can always be updated over time like that for any knowledge-based system [2]. Some of the rules used are: (1) An expression (or subexpression) should not end with any operator. (2) Horizontal operators do not have superscript or subscript expressions. (3) Superscript and subscript expressions of NH operators should not have matrix-like expressions, except determinants (as determinants are logically scalars). (4) Elements of a matrix do not have operators alone or they do not have equations.

Generation of Encoding Form: After errors in the ME tree structure are rectified by validation module, encoding form like LATEX, MathML etc., can be generated by traversing the tree. In our approach, LATEX code is generated by maintaining a mapping table that maps a given symbol to its LATEX encoded symbol. The algorithm to generate encoding form is based on pre-order traversal on the ME tree: (1) Inspect the label of the current node (which is head of tree initially) and generate its encoded symbol. (2) Encode its northern and southern subtrees recursively. (3) Inspect EOR and EOE fields and if they are true, add row and element delimiters (\\ and & for LATEX) accordingly. (4) Move to the next baseline symbol and repeat the first three steps.

An ME image is shown in Fig. 7(a). For this ME, delimiters are taken as MLME ones (determinants) with five rows and two columns by the symbol formation process and its LATEX output (without validation) is shown in Fig. 7(b). Its regenerated ME is also shown in Fig. 7(c) for better understandability. Validation module finds an equation in the first element of the last row ($x = a$) of the determinant and gives feedback to the symbol formation module that it is not a determinant. It is considered by symbol formation module and the remaining procedure is repeated to get correct tree structure. Its LATEX output as well as its regenerated ME are shown in Figs. 7(d) and (e) respectively.

Table 1. Summary of results on our PACME database of 829 MEs

ME type	Field	Number of MEs	Without Validation		With Validation	
			Number of correctly captured structures	Accuracy (%)	Number of correctly captured structures	Accuracy (%)
Non-MLMEs	Algebra	251	189	75.3	212	84.4
	Trigonometry	116	108	93.1	112	96.6
	Geometry	17	13	76.5	17	100
	Calculus	374	321	85.8	352	94.1
MLMEs	Matrices	62	50	80.6	53	85.4
	Enumerated functions	9	7	77.8	8	88.9
	Grand Total	829	688	82.9	754	90.9

6   Experimental Results and Discussion

For our experimentation, we have created a database of 829 ME images called
PACME database [10] that are collected from various mathematical books [20].
Those MEs cover different fields of mathematics like Algebra, Trigonometry, Geometry, Calculus etc., and range from simple to complex formulae and equations.
Each ME image in the database is binarized and CCs are extracted. Labels of
these CCs are manually stored in a file. Our approach to ME structural analysis has been implemented in C++. As mentioned in Section 3, for a given ME
image, proposed approach binarizes it, extracts the CCs and uses its corresponding file to label those CCs. After CC labelling, it performs the task of structural
analysis. Results on our PACME database of 829 MEs are summarized in Table
1. In that table, percentage accuracies without and with structure validation
are shown separately (to give a complete picture of our database) for each ME
type (MLMEs and Non-MLMEs) and field in that type. Accuracy is the ratio
of the number of correctly captured ME structures to the total number of MEs.
The above experimentation shows that validation process (based on our tree
structure) has given 8% improvement in accuracy.

7   Conclusions and Future Directions

In this paper, we have proposed a ternary tree based representation for structural
analysis of printed MEs. We have shown that the proposed tree is simple, complete and spatially efficient. It can be used to represent other similar structures
like chemical equations and also in applications like ME retrieval. Generated
tree structure is validated using domain knowledge and error feedback is used
to automatically correct the errors. Proposed approach that is based on an expert system incorporates an intelligent mechanism in the process. Our approach
has been tested on a database of 829 ME images and experimental results are
reported on them. In future, semantic knowledge can be incorporated into validation process to improve error detection and correction.

References

1. Álvaro, F., Sánchez, J.A., Benedí, J.M.: Classification of on-line mathematical symbols with hybrid features and recurrent neural networks. In: International Conference on Document Analysis and Recognition (ICDAR), pp. 1012–1016 (2013)

2. Buchanan, B.G., Shortliffe, E.H.: Rule Based Expert Systems: The Mycin Experiments of the Stanford Heuristic Programming Project. Addison-Wesley (1984)
3. Chan, K.F., Yeung, D.Y.: An efficient syntactic approach to structural analysis of on-line handwritten mathematical expressions. Pattern Recognition 33(3), 375–384 (2000)
4. Cormen, T.H., Leiserson, C.E., Rivest, R.L.: Introduction to Algorithms. The MIT Press and McGraw-Hill Book Company (1989)
5. Eto, Y., Suzuki, M.: Mathematical formula recognition using virtual link network. In: ICDAR 2001, pp. 762–767. IEEE Computer Society, Washington, DC (2001)
6. Garain, U., Chaudhuri, B.B.: Recognition of online handwritten mathematical expressions. IEEE Transactions on Systems, Man, and Cybernetics, Part B: Cybernetics 34(6), 2366–2376 (2004)
7. Gonzalez, R.C., Woods, R.E.: Digital Image Processing, 2nd edn. Pearson Education Indian Reprint (2003)
8. Lee, H.J., Wang, J.S.: Design of a mathematical expression understanding system. Pattern Recognition Letters 18(3), 289–298 (1997)
9. MacLean, S., Labahn, G.: A new approach for recognizing handwritten mathematics using relational grammars and fuzzy sets. IJDAR 16(2), 139–163 (2013)
10. PACME: Printed Mathematical Expression Image Database (2010), http://dcis.uohyd.ernet.in/~pavanp/mathocr/PrintedMEs.zip
11. Pavan Kumar, P., Agarwal, A., Bhagvati, C.: A rule-based approach to form mathematical symbols in printed mathematical expressions. In: Sombattheera, C., Agarwal, A., Udgata, S.K., Lavangnananda, K. (eds.) MIWAI 2011. LNCS (LNAI), vol. 7080, pp. 181–192. Springer, Heidelberg (2011)
12. Pavan Kumar, P., Agarwal, A., Bhagvati, C.: A structure based approach for mathematical expression retrieval. In: Sombattheera, C., Loi, N.K., Wankar, R., Quan, T. (eds.) MIWAI 2012. LNCS (LNAI), vol. 7694, pp. 23–34. Springer, Heidelberg (2012)
13. Pavan Kumar, P., Agarwal, A., Bhagvati, C.: A string matching based algorithm for performance evaluation of mathematical expression recognition. Sadhana 39(1), 63–79 (2014)
14. Suzuki, M., Tamari, F., Fukuda, R., Uchida, S., Kanahori, T.: Infty-an integrated OCR system for mathematical documents. In: Proceedings of ACM Symposium on Document Engineering 2003, pp. 95–104. ACM Press (2003)
15. Tapia, E., Rojas, R.: Recognition of on-line handwritten mathematical expressions using a minimum spanning tree construction and symbol dominance. In: Lladós, J., Kwon, Y.-B. (eds.) GREC 2003. LNCS, vol. 3088, pp. 329–340. Springer, Heidelberg (2004)
16. Tian, X., Fan, H.: Structural analysis based on baseline in printed mathematical expressions. In: PDCAT 2005, pp. 787–790 (2005)
17. Zanibbi, R., Blostein, D., Cordy, J.R.: Directions in recognizing tabular structures of handwritten mathematics notation. In: Proceedings of IAPR International Workshop on Graphics Recognition (2001)
18. Zanibbi, R., Blostein, D.: Recognition and retrieval of mathematical expressions. IJDAR 15(4), 331–357 (2012)
19. Zanibbi, R., Blostein, D., Cordy, J.R.: Recognizing mathematical expressions using tree transformation. IEEE Transactions on PAMI 24(11), 1455–1467 (2002)
20. Zwillinger, D.: CRC Standard Mathematical Tables and Formulae, 30th edn. CRC Press, Boca Raton (1996)

A New Preprocessor to Fuzzy c-Means Algorithm

Raveen S., P.S.V.S. Sai Prasad, and Raghavendra Rao Chillarige

School of Computer and Information Sciences
University of Hyderabad
Hyderabad, Telangana - 500046, India
raveenneevar@gmail.com, {saics,crrcs}@uohyd.ernet.in

Abstract. The fuzzy clustering scenario resulting from Fuzzy c-Means Algorithm (FCM) is highly sensitive to input parameters, number of clusters c and randomly initialized fuzzy membership matrix. Traditionally, the optimal fuzzy clustering scenario is arrived by exhaustive, repetitive invocations of FCM for different c values. In this paper, a new preprocessor based on Simplified Fuzzy Min-Max Neural Network (FMMNN) is proposed for FCM. The new preprocessor results in an estimation of the number of clusters in a given dataset and an initialization for fuzzy membership matrix. FCM algorithm with embeded preprocessor is named as FCMPre. Comparative experimental results of FCMPre with FCM based on benchmark datasets, empirically established that FCMPre discovers optimal (or near optimal) fuzzy clustering scenarios without exhaustive invocations of FCM along with obtaining significant computational gains.

Keywords: Soft Clustering, Fuzzy Clustering, Fuzzy c-Means Algorithm, Fuzzy Min-Max Neural Network.

1 Introduction

Clustering is a most important technique in data mining, which is used to discover groups and identify interesing distributions and patterns in the underlying data. Clustering divides data into different groups such that the data items belonging to the same groups are more similar than data items belonging to different groups. Clustering is widely used in various of applications. Some of the areas of application are business, biology, spatial data analysis, web mining etc. Clustering schema can be effectively used for data reduction, hypothesis generation, hypothesis testing and prediction based on groups. The clustering schema can be classified into different types based on different criteria. Based on the criteria of presence or absence of overlapping between clusters, clustering schema is classified into soft clustering, hard clustering respectively.

Soft clustering gives apt knowledge representation of real world scenarios as a data item belongs to more than one cluster (group) in several real world applications. Fuzzy clustering is an important soft clustering methodology based on the principles of fuzzy sets. In fuzzy clustering, a data item's belonging to a cluster is defined by a fuzzy membership function.

M.N. Murty et al. (Eds.): MIWAI 2014, LNAI 8875, pp. 124–135, 2014.

Fuzzy c-Means algorithm (FCM), is an important and widely used fuzzy clustering algorithm, was proposed by J. C. Bezdek in 1984 [16]. FCM is highly sensitive to input parameters, number of clusters c and randomly initialized fuzzy membership matrix. Hence in practice, repeated experiments need to be conducted using FCM for varying number of clusters, for different random initialization of fuzzy membership matrix to arrive at an optimal fuzzy clustering scenario. In this work, a novel preprocessor for FCM is proposed based on Simplified Fuzzy Min-Max Neural Network Algorithm (SFMMNN). The proposed preprocessor results in an estimation of the number of clusters c in a given dataset and an initialization of the fuzzy membership matrix. FCM algorithm with embedded preprocessor is named as FCMPre. The performance of FCMPre is analyzed based on comparative experimental analysis with FCM conducted on benchmark datasets.

Section 2 details the literature review of applications of FCM and improvements to FCM. FCM algorithm is described in Section 3. The proposed preprocessor to FCM and FCMPre algorithm is given in Section 4. Comparative experimental results and analysis is provided in Section 5.

2 Related Work

After its invention in 1984, FCM is applied in different application domains such as image processing, biology, bussiness etc. Several improvements to FCM are proposed in the literature. A review of applications and improvements to FCM is presented in this section.

W. E. Philips proposed application of Fuzzy c-Means segmentation technique for tissue differentiation in {MR} images of a hemorrhagic glioblastoma multiforme [8]. Demble proposed Fuzzy c-Means method for clustering microarray data [10]. A.B. Goktepe proposed Soil clustering by Fuzzy c-Means algorithm [6]. Keh-Shih Chuang proposed Fuzzy c-Means clustering with spatial information for image segmentation [7]. Li Xiang proposed an application of Fuzzy c-Means Clustering in Data Analysis of Metabolomics [11].

There are several approaches that have been proposed to improve the performance and efficiency of FCM. Hathaway proposed extension of Fuzzy c-Means Algorithm for Incomplete Datasets [9] containing missing values. Tai Wai Cheng proposed Fast Fuzzy clustering Algorithm (FFCM) [14] by incorporating a new procedure for initialization of cluster centroids replacing random initialization in FCM. Eschirich proposed Fast accurate fuzzy clustering through data reduction [12] by aggregation of similar data items into a single unit. Xizhao Wang improved Fuzzy c-Means clustering based on feature-weight learning [13]. This method uses proper learning techniques to find relevant features and avoids those features causing ambiguity. Al-Zoubi proposed a fast Fuzzy Clustering Algorithm [15]. The approach involves improving the performance of Fuzzy c-Means algorithm by eliminating unnecessary distance calculation. This can be done as comparing membership value with a threshold value and if it is less than the threshold value, the distance calculation is eliminated. The choice of number of clusters (c) for FCM is traditionally determined by optimizing validity index on results of several experiments conducted with diverse c values. Nguyen proposed a dynamic approach, in which adjustement to initial c value takes place along with cluster formation [18].

In this paper, an alternative approach for determining number of clusters and initial centroid vectors is proposed by using a preprocessor. The relevance of proposed preprocessor is empirically established by comparison with traditional approach. The next section describes FCM in detail.

3   Fuzzy c Means

Fuzzy clustering is proposed by Dunn in 1974 [3] and is extended by J.C. Bezdek in 1984 [16] as Fuzzy c-means Algorithm. The basic idea of fuzzy clustering is to partition given dataset $X = \{X_1, X_2, ...X_n\}$ into c number of fuzzy partitions. The fuzzy membership matrix U represents the resulting fuzzy partition. $U = [u_{ij}]_{N \times c}$ is a matrix with dimensions $N \times c$, where u_{ij} represents membership value of ith data object into j^{th} cluster. The objective function for finding optimal clustering schema for FCM is based on the least mean square method. The objective function is described as

$$J_m(U, v) = \Sigma_{k=1}^{N} \Sigma_{i=1}^{c} (u_{ik})^m \|X_k - v_i\|_A^2 \tag{1}$$

The variables used in objective function are
dataset $X = \{X_1, X_2, X_3...., X_N\}$ where $\forall X_i \in R^p$ (p : number of dimensions),
c : number of clusters; where $2 \leq c \leq N$,
m : weighting exponent which is called as fuzzifier ($1 < m < \infty$,)
U : fuzzy membership matrix (Fuzzy c-partition)
$v = (v_1, v_2, v_3, ...v_c)$: cluster centroids,
$v_i = (v_{i1}, v_{i2}, ..., v_{ip})$: i^{th} cluster centroid,
$\| \|_A$: induced A-norm on R^p

The FCM algorithm minimizes U and v in each iteration such that $J_m(U, v)$ is minimized. It uses an alternate optimization technique such that in each iteration of FCM, v is changed keeping U as fixed and then U is changed keeping v as fixed. The FCM algorithm is described in [16] and is given here for completeness.
Fuzzy c-Means Algorithm
Input:- Set of data vectors : $X = \{X_1, X_2, ...X_N\}$, c : number of clusters, m : fuzzifier, $\| \|_A$: distance norm.
Output:-U : Fuzzy c-partition of X.
Method
(1) Initialize membership matrix $U^{(0)}$.
(2) Calculate centroid vector $v^{(i)}, k = 1, 2, ..., c$ using the equation.

$$v^{(i)} = \frac{\Sigma_{k=1}^{N} (u_{ik})^m x_k}{\Sigma_{k=1}^{N} (u_{ik})^m} \tag{2}$$

(3) Calculate updated fuzzy membership matrix $U^q = [u^q]$ using the equation.

$$u_{ik}^q = \left(\Sigma_{j=1}^c \left(\frac{d_{ik}}{d_{jk}} \right)^{\frac{2}{(m-1)}} \right)^{-1} \tag{3}$$

where d_{ik} represents the distance measure of kth object to the centre vector of ith cluster.

(4) Compare $U^{(q)}$ to $U^{(q-1)}$. $if \|U^{(q)} - U^{(q-1)}\| < \epsilon$,stop. otherwise set $q = q+1$ and continue to step (2) .

There are different validation techniques available in the literature [16] to evaluate the validity of the fuzzy clustering schema. Popular validity indices are Partition Coefficient (PC), Partition Entropy (PE). Both PC and PE suffer from the monotonic evolution tendency with c. The validity index MPC (Modification of the PC) [2] reduces the monotonic tendency. Hence MPC is used in the present work for validation of fuzzy clustering.

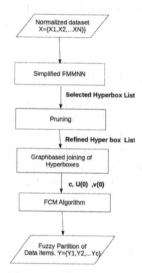

Fig. 1. Flow chart of FCMPre Algorithm

4 New Preprocessor to Fuzzy c-Means Algorithm

Fuzzy c-mean algorithm requires the number of clusters c as input parameter. Given a validation index V, the problem of obtaining optimal c-value is computationally highly intensive. The problem can be formally specified as follows. Assuming optimization requires maximization of V and N being the size of the dataset, the optimum c value is c^* where

$$V_{c^*} = max_{2 \leq c \leq (N-1)} V_c \tag{4}$$

The problem is further complicated owing to variability in behaviour of FCM algorithm due to random initialization of the fuzzy membership matrix. Usually in practice an experiment is repeated M times for a particular c value, for arriving at optimal fuzzy clustering. Hence for obtaining V^* in practice, requires $(N-2)*M$ invocations of FCM. In this section, a novel preprocessing technique is proposed for arriving at an estimation of c value and obtaining initial centroids which leads to a deterministic procedure for obtaining an initial fuzzy membership matrix so that an optimal or near optimal clustering configuration can be obtained without performing an exhaustive invocations of FCM.

The proposed approach involves applying a fast clustering algorithm for estimation of clusters followed by a graph based approach for merging of clusters (a decision guided by intended validation index). The resulting clusters determine the estimated c value and initial centroids which are adapted to FCM and if required iterations of FCM are continued till convergence.

The literature survey conducted has identified that Fuzzy Min-Max neural network (FMMNN), proposed by Simpson [17], is a fast algorithm for arriving at an estimation of clusters. In this work, a modification of FMMNN is used for the initial estimation of clusters in the proposed approach which is described in Section 4.1.

4.1 Fuzzy Min-Max Neural Network

Fuzzy min-max Neural Network for clustering is a single pass clustering algorithm proposed by Simpson [17]. FMMNN is a fast clustering algorithm for fuzzy clustering, which group the set of data vectors into different hyperboxes. Hyperboxes are the p-dimensional unit cubes which is associated with a membership function which describes to what membership degree the data vector belongs to a particular hyperbox fuzzy set. The membership degree ranges from 0 to 1 in which '0' means no membership and '1' means full membership. The FMMNN for clustering is discussed in Section 4.2.

4.2 Fuzzy Min-Max Clustering Algorithm

The first step of FMMNN is the initialization, which includes the formation of point hyperboxes (hyperbox with input data vector as min and max points) into an uncommitted set (UC) for all data vectors and a point hyperbox of the first data vector is introduced into a committed set (CS). After initialization, with each input pattern, a three step process (Expansion, Overlap test and Contraction) is applied.

In the literature, a simplified FMMNN for classification was proposed which uses only the initialization and expansion step of the original algorithm [4]. This results in overlapping hyperboxes in simplified FMMNN. Since the proposed preprocessor for FCM algorithm is meant for drawing an information about soft clustering and obtaining an initial fuzzy membership matrix for FCM, the overlapping of hyperboxes is allowed. Hence the concept of simplified FMMNN used for classification in [4] is adapted for proposed preprocessing strategy.

The Simplified Fuzzy Min Max Neural Network algorithm (SFMMNN) for clustering is given below.

Algorithm Simplified FMMNN (SFMMNN)
Input:- Set of data vectors, $X = \{X_1, X_2, \ldots\ldots, X_N\}$
Output:- Committed Set of Hyperboxes, $CS = \{B_1, \ldots\ldots B_l\}$
Method

1. Initialize θ, γ . Initialize uncommitted set (UC) with point hyperboxes for each data vector $X_h \in X$. Remove the first hyperbox from UC and include in committed set CS .
2. $\forall X_h \in \{X_2, \ldots X_N\}$
   (a) Select $B_j \in CS$ which satisfy equation
   $$\Sigma_{i=1}^{p}(max(w_{ji}, x_{hi}) - min(x_{hi}, v_{ji})) \leq p\theta$$

   (b) If B_j exists
      i. Update the max boundaries of B_j, which satisfies the expansion criteria.
      $$w_{ji}^{new} = max(w_{ji}^{old}, x_{hi})$$
      ii. Update the min boundaries of B_j as
      $$v_{ji}^{new} = min(v_{ji}^{old}, x_{hi})$$
      else
      i. Remove $B_h \in UC$ and include in CS .
3. Return CS

4.3   Fuzzy Min-Max Preprocesser for Fuzzy c-Means

The first step of the preprocessor is the application of simplified FMMNN. The resulting selected hyperboxes may be very high in number and some may contain very few data vetors and can mislead the recommendation of c value. A post pruning process is done to remove the irrelevant hyperboxes from the list, which does not contain at least $t\%$ of total input data vectors.

SFMMNN under the constraint of expansion parameter θ may represent a large cluster with multiple hyperboxes. In order to obtain the appropriate recommendation of c value, a graph based combining process is proposed. The graph based combining process checks whether the individual hyperboxes can be combined to give better result. The union of small hyperboxes into larger one will be based on the comparison of validity index (V_m) obtained for the current set of hyperboxes with an intended set of hyperboxes resulting from the union of two hyperboxes under consideration. The procedure for graph based approach for combining hyperboxes is given below.

Graph based approach for combining Hyperboxes

- A graph $G = (V, E)$ constructed from refined list of hyperboxes considering each hyperbox as nodes and placing an edge between overlapping hyperboxes
.
- Graph G is represented as adjacency list.
- $\forall v \in V$
  - $\forall u \in$ adjacencyList(v)
    1. Construct two fuzzy clustering scenarios.
       * **Scenario1:-** Fuzzy clustering resulting from the current set of hyperboxes.
       * **Scenario2:-** Fuzzy clustering resulting from collapsing u and v into single hyperbox along with remaining hyperboxes.
    2. Compute Validity index (V_m) for Scenario1 and Scenario2.
    3. if $V_m(Scenario2) > V_m(Scenario1)$
       * If true, then Collapse u and v into new hyperbox B_{new} and update the list.

After obtaining a selected list of hyperboxes using graph based combining, cluster centroids are calculated for these selected hyperboxes, based on the data vectors assigned to each hyperbox. The number of hyperboxes determines the recommended c value and their centroids form the initial centroid vectors of FCM algorithm. Based on centroid vectors, initial fuzzy membership matrix is computed. The execution of FCM is continued based on the newly obtained membership matrix.

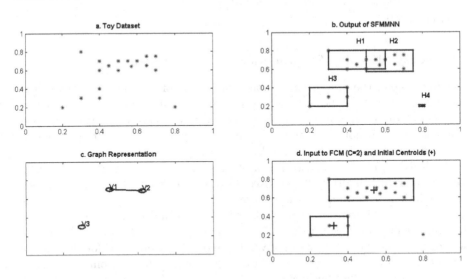

Fig. 2. Illustration of Preprocessor to FCM

Figure 2 illustrates the steps involved in the proposed preprocessor for FCM. A toy dataset with two features as given in Fig. 2-a is used for illustration. Applying SFMNN, resulted in four hyperboxes (see Fig. 2-b). During pruning step hyperbox H4 is dropped as it contains very few objects ($< t\%$ where t = 5). Figure 2-c depicts the graph constructed after pruning phase. Edge is included between vertices V1, V2 as the corresponding hyperboxes H1, H2 are overlapping. During graph based approach for combining, V1 and V2 got collapsed into a single vertex. In the resulting scenario of two hyperboxes, the centroid vectors are computed as an average of objects falling absolutely into hyperboxes. Hence the proposed preprocessor resulted in recommending number of clusters $c = 2$ and initial centroid vectors as depicted in Fig. 2-d.

FCM algorithm with inclusion of proposed preprocessing step is named as FCMPre algorithm. The flow chart of FCMPre algorithm is given in Fig. 1

5 Experiments and Results

5.1 Experimental Design

Code for FCM and FCMPre are developed in Java environments. Experiments are conducted on benchmark datasets from UCI Machine Learning Repository. The objective of experimentation is for the empirical assesment of relevance of proposed preprocessor by comparing the results of FCMPre with optimal results obtained from FCM using exhaustive approach described in Section 4.

Table 1. Description of datasets used for Experimental study

Name	Number of data vectors	Dimension
Aggregation	788	2
Ecoli	336	7
Glass	214	7
Iris	150	4
Yeast	1484	8

The problem of obtaining optimal c^* as described in Section 4 using FCM needs applying FCM for a repeated number of times, for varying values of c from 2 to $N - 1$. In our experimental result FCM is applied repeatedly for 10 times for varying c values from 2 to 25. The m value is fixed at 2.0 as it is the recommended value from literature [1]. The number of iterations and time taken to converge is computed in each invocation of FCM. The performance of FCMPre and FCM are validated on the basis of MPC Coefficient as the validation measure.

FCMPre algorithm requires SFMMNN algorithm as a preprocessing algorithm. SFMMNN requires specification of θ and γ. The recommended value of γ is 4.0 [5] and the same is used in experiments. The behaviour of SFMMNN is greatly influenced by value of θ. It is expected that different values of θ can result in different number of selected hyperboxes, which after pruning and graph based combining hyperboxes may result in different initial configuration (number of centers, initial centroid vectors) for FCM invocation as part of FCMPre algorithm. For a particular θ, γ values, repeated invocations of SFMMNN result in same initial configuration for FCM making repeated execution of FCMPre for a particular θ, γ not required. Hence arriving at optimal clustering configuration using FCMPre is taken as the best clustering configuration resulted from applying FCMPre for θ = 0.2, 0.3, 0.4, 0.5.

From now onwards "FCM experiment" represents applying FCM repeatedly for 10 times for c values from 2 to 25. "FCMPre experiment" represents applying FCMPre for θ values 0.2, 0.3, 0.4, 0.5.

5.2   Results

In FCM experiments, the maximum limit for c is fixed at 25, because all these datasets contain at most 10 clusters as per the information given from UCI Machine Learning Repository. The maximum value of c for yeast dataset is fixed at 23 due to its behaviour of taking too many iterations to converge for high values of c and the validity measures obtained are inferior. For each dataset, for repeated experiments that are done for different values of c, the best validity measure obtained is taken into consideration for further analysis. In case, if two executions resulted in the same value of MPC Coefficient, the one with less number of iterations is taken as best configuration for that c value.

The FCMPre is executed on these five datasets for different values of θ. The range of θ varies from 0.2 to 0.5.

5.3   Analysis

The experimental results reestablished the sensitivity of FCM for different values of c, for different random initialization of fuzzy membership matrix even for the same c.

The primary objective of FCMPre is to obtain optimal fuzzy clustering configuration without conducting exhaustive FCM experiments. Figure 3 and Fig. 4 depict the MPC validation index obtained for FCM (best value for each c value), FCMPre (θ = 0.2, 0.3, 0.4, 0.5) for e-coli and yeast dataset respectively. From the experimental results obtained, the primary inference is that except for yeast dataset, FCMPre experiment could achieve the optimal clustering configuration obtained by exhaustive FCM experiments. In Yeast dataset, FCMPre could obtain fourth best cluster configuration whose MPC value is close to the best MPC value obtained in FCM experiments.

The salient comparative results of FCM and FCMPre are summarized in Table 2. Table 2 reports computational time for " FCM experiment", "FCMPre experiment", the best θ value obtained by FCMPre experiment, best MPC values obtained by FCM, FCMPre experiments, best c values recommended by FCM ans FCMPre experiments. Based on Table 2, FCMPre resulted in obtaining the optimal configuration (best c , best MPC) as FCM for Aggregation, Ecoli, Glass, Iris datasets. A near optimal configuration is obtained by FCMPre for

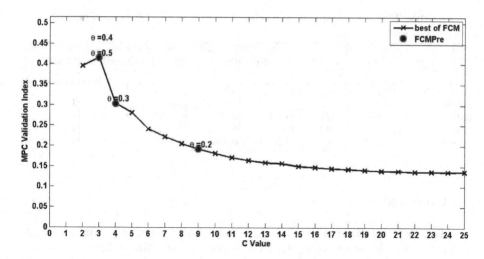

Fig. 3. Comparison of MPC validity measure for ecoli dataset

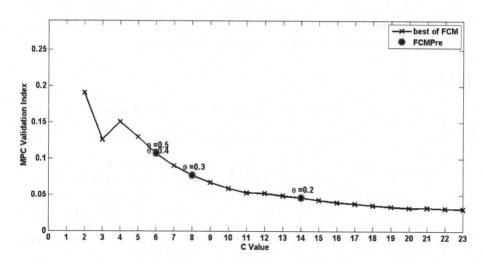

Fig. 4. Comparison of MPC validity measure for Yeast dataset

yeast dataset. These significant results are achieved by FCMPre with 99.5% and above computational gain over FCM.

Table 2. Salient comparative results of FCM and FCMPre experiments

Name	CT of FCM (sec)	CT of FCMPre (sec)	Best FCMPre	θ	Best MPC (FCM)	Best MPC (FCMPre)	Best C FCM	Best C FCMPre
Aggregation	7057.07	2.89	0.4		0.5934	0.5934	4	4
Ecoli	2209.45	5.91	0.5/0.4		0.4159	0.4159	3	3
Glass	1121.1	0.377	0.3/0.4		0.6653	0.6653	2	2
Iris	263.03	0.392	0.5		0.7199	0.7199	2	2
Yeast	73756.6	358.9	0.4/0.5		0.1905	0.1073	2	6

6   Conclusion

In this paper, algorithm FCMPre is proposed for arriving at optimal or near optimal fuzzy clustering using the FCM algorithm avoiding exhaustive experiments for varying sizes of c. FCMPre experiment is computationally effective and obtained huge computational gain over FCM experiments. FCMPre experiments resulted in optimal or near optimal fuzzy clustering scenario compared to FCM experiments. The experimental results empirically validate the utility of FCMPre for obtaining reliable fuzzy clustering scenario.

References

1. Pal, N.R., Bezdek, J.C.: On Cluster Validity for the Fuzzy C-means Model. Trans. Fuz. Sys. 3, 370–379 (1995)
2. Wang, W., Zhang, Y.: On fuzzy cluster validity indices. Fuzzy Sets and Systems 158, 2095–2117 (2007)
3. Dunn, J.C.: A Fuzzy Relative of the ISODATA Process and Its Use in Detecting Compact Well-Separated Clusters. Cybernetics and Systems 3, 32–57 (1973)
4. Ma, D., Liu, J., Wang, Z.: The Pattern Classification Based on Fuzzy Min-max Neural Network with New Algorithm. In: Wang, J., Yen, G.G., Polycarpou, M.M. (eds.) ISNN 2012, Part II. LNCS, vol. 7368, pp. 1–9. Springer, Heidelberg (2012)
5. Simpson, P.K.: Fuzzy min-max neural networks. In: 1991 IEEE International Joint Conference on Neural Networks, vol. 2, pp. 1658–1659 (1991)
6. Goktepe, A.B., Altun, S., Sezer, A.: Soil clustering by Fuzzy c-Means algorithm 36, 691–697 (2005)
7. Chuang, K.-S., Tzeng, H.-L., Chen, S., Wu, J., Chen, T.-J.: Fuzzy c-means clustering with spatial information for image segmentation. Computerized Medical Imaging and Graphics 30, 9–15 (2006)
8. Phillips II, W.E., Velthuizen, R.P., Phuphanich, S., Hall, L.O., Clarke, L.P., Silbiger, M.L.: Application of Fuzzy c-Means segmentation technique for tissue differentiation in MR images of a hemorrhagic glioblastoma multiforme. Magnetic Resonance Imaging 13, 277–290 (1995)
9. Hathaway, R.J., Bezdek, J.C.: Fuzzy C-means Clustering of Incomplete Data. Trans. Sys. Man Cyber. Part B 31, 735–744 (2001)
10. Dembl, D., Kastner, P.: Fuzzy C-means method for clustering microarray data. Bioinformatics 19, 973–980 (2003)

11. Li, X., Lu, X., Tian, J., Gao, P., Kong, H., Xu, G.: Application of Fuzzy c-Means Clustering in Data Analysis of Metabolomics. Analytical Chemistry 81, 4468–4475 (2009)
12. Eschrich, S., Ke, J., Hall, L.O., Goldgof, D.B.: Fast accurate fuzzy clustering through data reduction. IEEE Transactions on Fuzzy Systems 11, 262–270 (2003)
13. Wang, X., Wang, Y., Wang, L.: Improving Fuzzy c-Means clustering based on feature-weight learning. Pattern Recognition Letters 25, 1123–1132 (2004)
14. Cheng, T.W., Goldgof, D.B., Hall, L.O.: Fast fuzzy clustering. Fuzzy Sets and Systems 93, 49–56 (1998)
15. Al-Zoubi, M.B., Hudaib, A., Al-Shboul, B.: A Fast Fuzzy Clustering Algorithm. In: Proceedings of the 6th WSEAS Int. Conf. on Artificial Intelligence, Knowledge Engineering and Data Bases, vol. 3, pp. 28–32 (2007)
16. Bezdek, J., Ehrlich, R., Full, W.: FCM: The Fuzzy c-Means clustering algorithm. Computers & Geosciences 10, 191–203 (1984)
17. Simpson, P.K.: Fuzzy min-max neural networks - Part 2: Clustering. IEEE Transactions on Fuzzy Systems 1 (1993)
18. Nguyen, D.T., Doan, H.: An Approach to Determine the Number of Clusters for Clustering Algorithms. In: Nguyen, N.-T., Hoang, K., Jędrzejowicz, P. (eds.) ICCCI 2012, Part I. LNCS (LNAI), vol. 7653, pp. 485–494. Springer, Heidelberg (2012)

Domain Specific Sentiment Dictionary for Opinion Mining of Vietnamese Text

Hong Nam Nguyen, Thanh Van Le, Hai Son Le, and Tran Vu Pham

Faculty of Computer Science and Engineering
Ho Chi Minh City University of Technology, VNU-HCM
No 268, Ly Thuong Kiet Street, District 10, Ho Chi Minh City, Vietnam
{nhnam,ltvan,lhson,t.v.pham}@cse.hcmut.edu.vn

Abstract. Knowing public opinions from subjective text messages vastly available on the Web is very useful for many different purposes. Technically, extracting efficiently and accurately the opinions from a huge amount of unstructured text messages is challenging. For English language, a common approach to this problem is using sentiment dictionaries. However, building a sentiment dictionary for less popular languages, such as Vietnamese, is difficult and time consuming. This paper proposes an approach to mining public opinions from Vietnamese text using a domain specific sentiment dictionary in order to improve the accuracy. The sentiment dictionary is built incrementally using statistical methods for a specific domain. The efficiency of the approach is demonstrated through an application which is built to extract public opinions on online products and services. Even though this approach is designed initially for Vietnamese text, we believe that it is also applicable to other languages.

Keywords: opinion mining, sentiment dictionary, sentiment analysis.

1 Introduction

Subjective text messages in forms of reviews, blog posts and comments in forums is now vastly available on the Web. With the increasing use of Internet accessible personal devices such as smartphones and tablets, the number of text messages keeps increasing. People could use these text messages to express their opinions about many things such as products, services, social events, or their interests. Knowing public opinions by extracting valuable information from these text messages is really beneficial to many different parties such as policy makers, businesses, and manufacturers. However, with the huge amount of text messages currently available on the Web, extracting efficiently and accurately the opinions from the written text is a challenging problem.

In recent studies, a common approach is classifying text messages into categories such as positive, negative, or neutral support to a subject matter [1–4]. Sentiment dictionaries are usually used for looking up sentiments of individual words. Sentiments of individual words are then aggregated to sentences, then paragraphs, and eventually to the whole message. This approach is simple to

M.N. Murty et al. (Eds.): MIWAI 2014, LNAI 8875, pp. 136–148, 2014.
© Springer International Publishing Switzerland 2014

implement, but it requires good sentiment dictionaries. There exist sentiment dictionaries in English, such as SentiWordNet [5] which is derived from Word-Net. There is also an attempt to build a sentiment dictionary for Vietnamese language by translating an English sentiment dictionary into Vietnamese [6]. However, because of the differences between the two languages, the accuracy of the Vietnamese version is limited.

In this paper, we propose an approach to mining public opinions from Viet-namese short text messages using a domain specific sentiment dictionary in order to improve the accuracy. The sentiment dictionary is built incrementally using statistical methods for a specific domain. In this study, we develop an applica-tion to extract public opinions on online products and services to experiment and evaluate the efficiency of the approach. Even though this approach is de-signed initially for Vietnamese text, we believe that it is also applicable to other languages.

This paper is organized into seven sections. The next section briefly discusses the related work. Then, Section 3 provides a summary on a dictionary based approach to opinion mining, which is the basis of the proposed method. The core contribution of this paper is described in Section 4, 5 and 6. These sections show in detail how the domain specific sentiment dictionary is built, the way it is used to mining opinions and the experiments have been done to evaluate the work. Section 7, the last section, concludes the paper with a couple of remarks.

2 Related Work

Sentiment analysis, also known as opinion mining, is an activity that focuses on analyzing people opinions, sentiments, attitudes, and emotions towards tar-get objects such as products (mobile phones, computers, etc.), services (hotels, restaurants, etc.), individuals (politicians), or topics (economic, culture, customs, etc.). The main purposes of sentiment analysis are to detect and extract sub-jective information in text documents for determining sentiments of the writers about some aspects or the overall contextual polarity (positive, negative, sup-portive, or neutral) expressed in the documents [7–10].

In opinion mining, the most considered problem is sentiment classification [1–4] which labels documents as positive, negative, or neutral. There are three main levels of sentiment classification: document level, sentence level, entity and aspect level. Sentiment words are commonly used as key factors to determine whether a piece of text or a document has positive or negative sentiment. For in-stance, *wonderful, perfect, amazing* are sentimentally positive words and *bad, aw-ful, terrible* are sentimentally negative words. Due to the importance of sentiment words within sentences, many approaches have proposed to generate sentiments of words which could be used to support sentiment classification of documents. There are three common approaches to generating sentiments of words: manual, dictionary-based and corpus-based approaches. The manual approach simply uses human knowledge to decide the sentiment of a word. Meanwhile, dictionary-based and corpus-based approaches automatically generate sentiments of words using dictionaries and corpuses respectively.

In 2013, Rao et al. [10] proposed an algorithm to build a word-level emotional dictionary automatically for social emotion detection. Their method was based on Emotion LDA (Latent Dirichlet Allocation) model (ELDA) which jointly modeled emotions and topics by LDA. In this method, word-level emotional dictionary was constructed based on maximum likelihood estimation and Jensen's inequality. As the emotional words were generated from real world news corpuses, the emotional dictionary contained topical common words. A pruning algorithm was proposed to refine the emotional dictionary by removing common unwanted words. At topic-level, LDA was used. Their study showed that ELDA together with their pruning algorithm could give good performance with certain setting. There has been approach using sentiment lexicons for classifying documents [11]. In this approach, a Machine Learning Senti-word Lexicon was built using Support Vector Machine (SVM) with the dataset containing reviews and comments from Amazon. Linguistic features of text was used during the learning. In Chinese, Wang et al. [8] proposed to use opinion lexicons for unsupervised opinion phrase extraction and rating mechanism. The assumption was that neighboring terms were usually modifiers of the opinion terms to describe their orientation and strength. Then for each opinion term, the neighboring terms were evaluated to decide the score. Finally, average score of the all opinion phrases was the final score of the post.

For Vietnamese text, Tien-Thanh Vu et al. (2011) [12] proposed an opinion mining model to extract sentiments from reviews on mobile phone products. Feature words and opinion words were extracted using some Vietnamese syntactic rules. Opinion orientation and summarization on features were determined by using VietSentiWordNet[1] (with 1017 words). In 2003, Ho et al. [6] proposed an approach to build Vietnamese WordNet based on the hypothesis that the inheritance systems of nominal concepts in Vietnamese and English languages are similar, at least in certain domains. Their work was focused mainly on creating a hierarchy of Vietnamese nouns, synonym sets and the links among them using English WordNet, bilingual Vietnamese-English and English-Vietnamese dictionaries, and Vietnamese dictionaries.

In addition, one of popular sentiment lexicon is SentiWordNet [5] which contains opinion information on terms extracted from the WordNet by using a semi-supervised learning method and made publicly available for research purposes. SentiWordNet provides a readily available database of term sentiment information for English language and could also be used as a reference to the process of building and extending opinion dictionaries for Vietnamese language.

3   Dictionary-Based Approach

Using a dictionary to extract sentiment words is one of the main approaches in opinion mining [7]. Result of this approach is an emotion word set containing opinion-bearing words (positive sentiment and negative sentiment words). Positive sentiment words are used to express some desired states and good qualities

---

[1] http://sourceforge.net/projects/vietsentiwordnet/

while negative sentiment ones are used to express some undesired states or bad qualities. The idea of using sentiment dictionaries in opinion mining is originated from the hypothesis that individual words can be considered as a unit of opinion information, and therefore may provide clues to the overall sentiment the document.

SentiWordNet[2] is a lexical resource of sentiment information for terms in English language designed to assist in opinion mining tasks. Each term in Senti-WordNet is associated with numerical scores for positive and negative sentiment information. In our approach, these scores were used to build a Vietnamese opinion dictionary. However, the number of terms defined in SentiWordNet are limited. We attempt to extend this dictionary by using WordNet score propagation algorithm[9], as described in Fig. 1 to measure the score of a word. In this figure:

- seed sets P (positive), N (negative) and M (neutral)
- score vector s_0 is initialized(all positive seed words get a value of $+1$, all negative seed words get a value of -1, and all other words a value of 0)
- λ is scaling factor ($\lambda < 1$)
- $syn(w_i)$, $ant(w_j)$ are synonym and antonym sets extracted from WordNet
- Matrix $A = (a_{ij})$ is simply a matrix that represents a directed

Seed-sets	Score vector s_0	Matrix A
P (pos.) N (neg.) M (obj.)	$s_0^i = \begin{cases} +1 \text{ if } w_i \in P \\ -1 \text{ if } w_i \in N \\ 0 \text{ otherwise} \end{cases}$	$a_{i,j} = \begin{cases} 1+\lambda \text{ if } i=j \\ +\lambda \text{ if } w_i \in syn(w_j) \text{ \& } w_i \notin M \\ -\lambda \text{ if } w_i \in ant(w_j) \text{ \& } w_i \notin M \\ 0 \text{ otherwise} \end{cases}$

Repeated $A*s_i$: $A*s_0 = s_1$; $A*s_1 = s_2$; ... sentiment scores from the final vector (and, for each item, change its final sign to its initial sign if the two differ)

Fig. 1. The WordNet score propagation algorithm

SentiWordnet is a lexical resource for opinion mining and hence is currently applied in many research projects for extracting sentiment words or building extended dictionaries. However, existing sentiment dictionaries have some limitations when applied to analyzing Vietnamese comments and reviews for mining user opinions. These dictionaries usually lack of many sentiment words, particularly in specific domains. In addition, the problem is worsen as word combination rules are very much different between English and Vietnamese. In Vietnamese, the overall opinion of a sentence is not only determined by single words but also by phrases of words. Furthermore, these dictionaries do not contains slang words.

---

[2] http://sentiwordnet.isti.cnr.it/

These slang words appear very often in Web comments and reviews. Therefore, in our approach, we propose a method for determining opinion orientation based on our extended dictionary for Vietnamese language, which is presented in detail in the next section.

4 Our Approach

In our proposed approach, there are two main tasks: build and enrich Vietnamese opinion dictionary; sentiment analyses.

4.1 Build Opinion Dictionary

In this section we describe the component of our system that build Vietnamese opinion dictionary based on SentiwordNet and enrich it based on WordNet and list of words extracted from reviewer comments.

General overview of the system enrich opinion dictionary is following Figure 4.1. The first step of our procedure is to choose list of English adjective, adverb and verb in that can be constitute the skeleton of building Vietnamese opinion dictionary from Sentiwordnet 3.0 [?] and to identify all meanings of those words. English-Vietnamese dictionary used to translate all of words in list to Vietnamese and created of each entry the set of meanings of the headword if it is one of three types (adjective, adverb and verb). There is noise in the result data set of entries English-Vietnamese because not the all English words exist in dictionary and each English word would be have one or more senses or do not clear meaning. As mentioned previously, the next step we need to clean noise in the set of entries. For that task we need a list that comprises almost all Vietnamese adjective, adverb and verb. We have extracted such a list from available monolingual dictionaries and Vietnamese-English dictionary. Basically that we have here is a Vietnamese opinion dictionary based on Sentiwordnet 3.0 [?], but it is not enough to use determining opinion of comment in Vietnamese.

Therefore the next task, we will enrich this dictionary by the popular opinion words, which extracted from data set comment. Comments product crawled from website will be standardized, such as *"ko tốt"* switched to *"không tốt"* (no-good). After that, we use WordSeg tool [?] to practice token segmenting and POS tagging.

Due to the characteristics of Vietnamese, a word might contain only one individual word (one morpheme) or a compound of two or more individual words (many morphemes). This tool identifies words and tokenizes sentences into separate tokens. Resulting tokenized documents are used for further analysis tasks.

The obtained result from above example was: *Máy/**N** khởi động/**V** rất/**R** chậm/**A**. (Machine (smart phone) boots very slow.)* in which **N** is a Noun, **V** is a Verb, **R** is adverb and **A** is adjective. Then, variations of the TF–IDF (term frequency - inverse document frequency) weighting scheme are used finding list of words along with their type, which often used in comments. In this case, words at the top of the list have a low TF-IDF weight will be chosen.

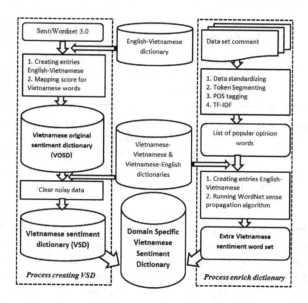

Fig. 2. System architecture of the proposed approach enrich Vietnamese opinion dictionary

Table 1. List of popular words (a = adjective, r = adverb)

POS	Word	POS	Word
a	đẹp *(beautiful)*	a	tốt *(good)*
a	sang trọng *(luxurious)*	a	dễ *(easy)*
a	mượt *(smooth)*	a	mạnh *(strong)*
a	đầy đủ *(complete)*	a	chắc chắn *(secure)*
r	không *(not)*	r	cũng (also)
a	tương đối *(relative)*	a	tiện lợi *(convenient)*

Table 1 depicts a tiny fraction of list of popular words of result.

We go through all words of list and find English words in WordNet [?], which has meaning match with their. After that we has list of English word and it used to build score for Vietnamese opinion word. The WordNet is used to grow sets of positive and negative terms. However, in our work we wish not only to create these sets, but also to weigh each member of the set with a confidence measure that represents how likely it is that the given word has the designated positive or negative sentiment. Thus, we use the WordNet score propagation algorithms [?] such as Figure 4.1 measure score of word.

The inputs to the algorithm are the three manually constructed seed sets that we denote as P (positive), N (negative), and M (neutral). Also provided as input are the synonym and antonym sets extracted from WordNet for arbitrary word w and denoted by $syn(w)$ and $ant(w)$ respectively.

Seed-sets	Score vector s_0	Matrix A
P (pos.) N (neg.) M (obj.)	$s_0^i = \begin{cases} +1 \text{ if } w_i \in P \\ -1 \text{ if } w_i \in N \\ 0 \text{ otherwise} \end{cases}$	$a_{i,j} = \begin{cases} 1+\lambda \text{ if } i=j \\ +\lambda \text{ if } w_i \in \text{syn}(w_j) \ \& \ w_i \notin M \\ -\lambda \text{ if } w_i \in \text{ant}(w_j) \ \& \ w_i \notin M \\ 0 \text{ otherwise} \end{cases}$

Repeated $A*s_i$: $A*s_0 = s_1$; $A*s_1 = s_2$; ...   sentiment scores from the final vector $\begin{pmatrix} \text{and, for each item, change its final} \\ \text{sign to its initial sign if the two differ} \end{pmatrix}$

Fig. 3. The WordNet score propagation algorithm

The algorithm begins by defining a score vector s^m that will encode sentiment word scores for every word in WordNet. This vector will be iteratively updated (each update indicated by the superscript). s^0 is initialized so that all positive seed words get a value of +1, all negative seed words get a value of -1, and all other words a value of 0. Next, we choose a scaling factor $\lambda < 1$ to help define an adjacency matrix for the set of all words wi in the WordNet lexicon $\mathbf{A} = (a_{ij})$. Where A is simply a matrix that represents a directed, edge-weighted semantic graph where neighbouring nodes are synonyms or antonyms and are not part of the predefined neutral set – the latter being necessary to stop the propagation of sentiment through neutral words.

We then propagate the sentiment scores over the graph via repeated multiplication of $\mathbf{A}$ against score vectors s^m, augmented with a sign-correction function for the seed words to compensate for relations which are less meaningful in the context of reviews:

$$s^m := sign - correct(As^{m-1})$$

Here, the function t = sign-correct(s) maintains $|t_i| = |s_i| \forall i$, ensures that $sign(t_i) = s_i^0$ for all seed words w_i , and preserves the sign of all other words, and maintain its sign at each of the m iteration.

The final score vector $\mathbf{s}$ is derived by logarithmically scaling $\mathbf{s}^M$

$$\mathbf{s}_i = \begin{cases} log(|s_i^M|) * sign(|s_i^M|) \ if |s_i^M| > 1 \\ 0 \qquad\qquad\qquad\qquad\quad otherwise \end{cases}$$

We use the decaying parameter λ to limit the magnitude of scores that are far away from seeds in the graph. In our experiments we used $\lambda = 0.2$ and ran for M = 5 iterations. We also scaled score in limit from -1 to 1 like SentiwordNet.

Running the algorithm resulted in an expanded sentiment lexicon, some of which are shown in Table 2 with their final scores.

In addition, we are analyzing emotions of reviewer comments, which write on online website, so that the structure of spoken language is used more than the structure of written language in their comment. Especially the slang commonly used and the also contributed significantly to identification of user comments. For example: *Điện thoại củ chuối. (banana telephone).* The word "củ chuối"

Table 2. Example terms from our induced sentiment lexicon, along with their scores(a = adjective, r = adverb)

POS	Word	Score
a	tạm	0.256
v	bắt mắt	0.125
v	hạn chế	-0.360
v	ổn định	0.3
v	nhậy	0.42
a	rẻ	0.45
a	dở	-0.40
v	dơ	-0.3
a	tiện	0.25

(banana) is mean not good and they do not like the telephone $\longrightarrow$ *negative opinion*. Therefore, in enrich Vietnamese opinion dictionary task they have been handled and added to dictionary.

4.2 Sentiment Analysis

Unigram. In Vietnamese sentence, the most of the elementary representation of an opinion word is the signle word (unigram) such as *tốt(good)*, *đẹp(beautiful)*, etc.. For example the sentence *Diện thoại này giá không quá đắt. (This phone price is not too expensive)*, here unigrams formed are *'Diện thoại'(phone)*, *'giá' (price)*, *'không'(not)*, *'quá'(too)*, *'đắt'(expensive)* but they does not carry sufficient information for opinion because in unigram model do no consider the relationship between words in sentence. To avoid this problem, word phrase is used to identify opinion orientation in our approach.

Phrase Analysis. In our approach, reviewer comment **C** is content one or more sentence. The sentence **S** is represented with finite set of phrases F $=\{f_1,...,f_n\}$. Each $f_i \in$ F can be expressed with any one of a finite set of opinion words OpW $=\{opw_1,..., opw_n\}$ and related words RW $= \{rw_1,...,rw_n\}$ (the words contribute to the changing orientation of the emotional term.

We also have basic rule of opinion changing:

1. Neg implies a negative opinion (denoted by Negative) and Pos implies a positive opinion (denoted by Positive) in a phrase. The effect of negations can be represented as well: *Negation Neg* $\longrightarrow$ *Positive, Negation Pos* $\longrightarrow$ *Negative.* The rules state that negated opinion phrases take their opposite orientations in a sentence.

2. Decreasing or increasing the quantities associated with some related word may change the orientations of the opinions.

Decreased Neg $\longrightarrow$ Positive
Decreased Pos $\longrightarrow$ Negative
Increased Neg $\longrightarrow$ Strong Negative (Negative)
Increased Pos $\longrightarrow$ Strong Positive (Positive)

The opinion orientation on each phrase f_i in s determined by the following opinion function:

$$score(f_i, s) = \Sigma opw_j, rw_k$$

The opinion orientation on sentence s_i in c and c_i determined by the following opinion functions:

$$score(s_i) = \Sigma score(f_i); score(c_i) = \Sigma score(s_i)$$

4.3   Experiment

This section provides a brief details of the datasets used by us in our experiments and the result of our approach.

Data Pre-processing. We have developed a tool used to collect comments on the online sales website. Sentiment dataset contains reviews from commercial website. We considered 4000 comments for training; 1000 positive and 1000 negative reviews randomly sampled from 20083 comments for mobile product were be crawled on different websites for testing. The data set of comment do not contain long sentences and do not have a lot of complex sentences. The following data preparation steps are performed on this dataset before it is used.

1.*Data Standardizing*: The reviewers often use a combination of standard spelling, apparently accidental mistakes, and interjections in their reviews. Therefor we need correct problems and make sentence in this step before go to the next step.

2.*Token Segmenting and POS Tagging*: Due to the characteristics of Vietnamese, a word might contain only one individual word (one morpheme) or a compound of two or more individual words (many morphemes). This step identifies words and tokenizes sentences into separate tokens (by vnTokenizer tool). The POS tagger (by VietTagger tools) uses tokenized documents as input and assigns a POS label for all tokens.

3. *Segmenting sentence to phrase*: In phase, segmenting sentences into chunks helps determine grammatical roles of elements in sentence because noun phrases, adjective phrases and verb phrases are used as majority units of the patterns in our approach. The output we have a set of phrase with each phrases contain one or more words and their type.

5   Experiments

5.1   Data Pre-processing

For experimental purpose, 20083 comments were crawled from the Web. 1000 positive and 1000 negative comments are randomly selected this set to train the application. Most of the comments consisted of short and simple sentences. The following data preparation steps were performed on this dataset before training:

1. *Data Standardizing*: The reviewers often used a combination of standard spelling, apparently accidental mistakes, and interjections in their reviews. Therefore, it was necessary to convert them to the formal form of Vietnamese. ("j" ↭ "gì", "ko" ↭ "không",...).
2. *Tokenizing and POS Tagging*: Due to the characteristics of Vietnamese, a word might contain one morpheme or a compound of two or many words (many morpheme). This step identified words and tokenized sentences into separate tokens. The POS tagger used tokenized documents as input and assigned POS labels to all tokens.
3. *Segmenting sentences to phrases*: This step segmented a sentence into chunks for determining grammatical roles of elements in the sentence because noun phrases, adjective phrases and verb phrases were used as major units of the patterns. The output of this step was a set of phrases where each element contained one or more words and their type (Adjective, Adverb or Verb).

5.2   Experimental Results

Different experiments were carried out with 5 different datasets. Each dataset consisted of 200 comments randomly sampled from 20083 comments (100 positive and 100 negative comments) for mobile products. In these experiments, the domain specific Vietnamese dictionary was used to determine sentiment orientation of the comments. In order to measure the efficiency of the domain specific Vietnamese sentiment dictionary, other dictionary such as VietSentiWordNet (version 1.0) and VOSD (the intermediate dictionary) were also used. The results of using different dictionaries were cross compared with each others.

The following formula were used to measure the accuracy of determining comment sentiments resulted from our approach:

$$acc = \frac{TP + TN}{P + N} \quad . \tag{1}$$

where:

- TP: number of true positive comments,
- TN: number of true negative comments,
- P is the number of positive comments,
- N is the number of negative comments.

Table 3. Results of determining orientation opinion with different sentiment dictionaries

Data set	1	2	3	4	5	average
VietSentiWordNet ver.1.0	61%	53.5%	57%	58%	57.29%	57.36%
VOSD	73.86%	77.5%	67.83%	74.24%	77.89%	74.26%
Extended dictionary	**82.92%**	**84.54%**	**79.98%**	**83.75%**	**81.58%**	**82.55%**

Experiment results showed that our domain specific Vietnamese sentiment dictionary produced higher accuracy than the VietSentiWordNet and VOSD. The best accuracy the domain specific dictionary could give was 84.54%, whereas VietSentiWordNet and the VOSD was only at 61% and 77.89%, respectively (see Table 3). Some of the results with different approaches on three comments extracted from the test data set are shown below:

1. *máy hơi nóng, pin nhanh hết (the phone is hot, spend battery very fast)* ▶ negative opinion
2. *pin cùi có 1000 mAh . nhanh hết pin so với cái màn hình to (battery is very bad. spending battery very fast with a big display)* ▶ negative opinion
3. *nghe gọi rất ok. phong cách thời trang (listening calling is so ok. fashion style)* ▶ positive opinion

Table 4 showed results of intermediate results during the processing. It shows the tokens extracted from the three comments using different dictionaries and their respective sentiment scores.

Table 4. Example of opinion mining on sample data with different dictionaries (A: adjective, R: adverb, V: verb, AP: adjective phrase, VP: verb phrase, NP: noun phrase)

No.⇓	VietSentiWordNet	VOSD	domain specific dictionary
1	nóng (A)--→ positive *nóng: (hot)*	nóng (A)--→ negative	hơi nóng (AP)--→ negative *(hơi nóng: rather hot)*
	nhanh (A)--→ positive *(nhanh: fast)*	nhanh (A)--→ positive	nhanh hết (VP)--→ negative *(nhanh hết: quickly out)*
	sentence ⟹ **positive**	sentence ⟹ **positive**	sentence⟹ **negative**
2	nhanh (A) --→ positive	có(V) --→ positive *(có: have)*	cùi có (VP)--→ negative *(cùi (A): something is very bad)*
		nhanh (A)--→ positive	nhanh hết (VP)--→negative
		to(A)--→ positive *(to: big)*	cái màn hình to (NP) --→ positive *(the screen is big)*
	sentence ⟹ **positive**	sentence ⟹ **positive**	sentence ⟹ **negative**
3	not found sentiment word	gọi (V)--→ positive *(gọi: call)*	nghe gọi rất okie (VP) --→positive
		rất (R)--→ positive *(rất: very)*	phong cách thời trang (NP) --→positive *(fashion style)*
		thời trang(N)--→positive	
	sentence ⟹ **unknown**	sentence ⟹ **positive**	sentence⟹ **positive**

Table 4 showed that using VietSentiWordNet (the first column) gave incorrect result on all three sentences. In the first and second sentences the word *"nóng (hot)"*, *"nhanh (fast)"*, *"to (big)"* have positive meaning. However, in this case, it gave negative meaning. With the third sentence, it could not find opinion words because this dictionary did not cover the all opinion words. Therefore, the result was unknown. With VOSD (the second column), the results were also incorrect for the same reason. The case of using domain specific dictionary, it produced correct results for all three sentences. The reason was that this dictionary had been enriched with opinion words.

It was common that comments often spanned over multiple sentences. It was specially difficult to determine the sentiment level of the whole comment if there were sentences with opposite sentiments. Our solution was that the sentiment of each sentence was calculated, then, sentiment of all sentences of the comments were aggregated together. The aggregated value was used to decide the sentiment of the whole comment. For example *"Giá còn cao. Các tính năng kỹ thuật tốt. Kích thước hơi lớn. Dễ sử dụng."* *(The price is high. The technical features are good. The size is so large. Easy to use.).* In this reviewer's comment, the first and the third sentences show negative polarity; the second and fourth sentences show positive polarity. Our system analyzed this comment and got a positive score because "tốt" *(good)* has a high positive score.

An observation we have noticed that our model performs well when reviews were in a specific domain (i.e. smart phone, computer reviews). The performance drops if the reviews come from different domains. In Vietnamese language, there are some cases that one opinion word could be interpreted by different meanings. There may also be different ways to express one opinion. A word "lâu" *(long)* is a positive sentiment if it refers to the battery but becomes negative sentiment when we say something about the waiting time in a restaurant. This is the reason why domain specific sentiment dictionaries are necessary in these cases to increase the accuracy of mining results.

6   Conclusion

This paper introduces an approach for mining public opinions from Vietnamese text using a custom built domain specific sentiment dictionary. The experiments show that with this approach it is able to detect sentiment words in short comment sentences although it is represented by spoken language. Processing public comments for opinion mining is a difficult task because opinions are commonly expressed in natural languages. The experiments also show that combining VSD with phrase analysis gives better results in determining opinion orientation on product reviews. Our solution is very promising even when a comment spans in multiple sentences and in negation form. For increasing the performance of our work, we must continue collecting slang words, acronyms, specific words on domain as much as possible for getting better results in mining public opinions on websites, forums or social networks. Even though our approach is initially designed for Vietnamese text, we believe that it is also applicable to other languages.

Acknowledgments. We would like to thank Toyo Information Systems Incorporation[3] for supporting and funding this research.

References

1. Fan, M., Wu, G.: Aspect opinion mining on customer reviews. In: Jiang, L. (ed.) ICCE2011. AISC, vol. 112, pp. 27–33. Springer, Heidelberg (2011)
2. Saif, H., He, Y., Alani, H.: Semantic sentiment analysis of twitter. In: Cudré-Mauroux, P., et al. (eds.) ISWC 2012, Part I. LNCS, vol. 7649, pp. 508–524. Springer, Heidelberg (2012)
3. Kumar, A., Sebastian, T.M.: Sentiment analysis: A perspective on its past, present and future. International Journal of Intelligent Systems and Applications (IJISA) 4(10), 1–14 (2012)
4. Pang, B., Lee, L., Vaithyanathan, S.: Thumbs up?: Sentiment classification using machine learning techniques. In: Proceedings of the ACL 2002 Conference on Empirical Methods in Natural Language Processing, EMNLP 2002, vol. 10, pp. 79–86. Association for Computational Linguistics, Stroudsburg (2002)
5. Esuli, A., Sebastiani, F.: Sentiwordnet: A publicly available lexical resource for opinion mining. In: Proceedings of the 5th Conference on Language Resources and Evaluation (LREC 2006), pp. 417–422 (2006)
6. Duc, H.N., Thao, N.T.: Towards building a wordnet for vietnamese (2003)
7. Liu, B.: Sentiment Analysis and Opinion Mining. In: Synthesis Lectures on Human Language Technologies. Morgan & Claypool Publishers (May 2012)
8. Wang, J.H., Lee, C.C.: Unsupervised opinion phrase extraction and rating in Chinese blog posts. In: SocialCom/PASSAT, Boston, MA, pp. 820–823. IEEE (2011)
9. Blair-goldensohn, S., Neylon, T., Hannan, K., Reis, G.A., Mcdonald, R., Reynar, J.: Building a sentiment summarizer for local service reviews. In: NLP in the Information Explosion Era, Beijing, China (2008)
10. Rao, Y., Lei, J., Wenyin, L., Li, Q., Chen, M.: Building emotional dictionary for sentiment analysis of online news. World Wide Web 17, 732–742 (2013)
11. Hamouda, A., Marei, M., Rohaim, M.: Building machine learning based senti-word lexicon for sentiment analysis. Journal of Advances in Information Technology (JAIT) 2(4), 199–203 (2011)
12. Vu, T.-T., Pham, H.-T., Luu, C.-T., Ha, Q.-T.: A feature-based opinion mining model on product reviews in vietnamese. In: Katarzyniak, R., Chiu, T.-F., Hong, C.-F., Nguyen, N.T. (eds.) Semantic Methods. SCI, vol. 381, pp. 23–33. Springer, Heidelberg (2011)

---

[3] http://www.tis.com

Support Vector–Quantile Regression Random Forest Hybrid for Regression Problems

Ravi Vadlamani[1,*] and Anurag Sharma[2]

[1] Center of Excellence in CRM and Analytics
Institute for Development and Research in Banking Technology,
Castle Hill, Masab Tank, Hyderabad 500057 (A P) India
[2] Department of Mathematics & Statistics,
Indian Institute of Technology, Kanpur 208016 (U P) India
vravi@idrbt.ac.in, sharmaan@iitk.ac.in

Abstract. In this paper we propose a novel support vector based soft computing technique which can be applied to solve regression problems. Proposed hybrid outperforms previously known techniques in literature in terms of accuracy of prediction and time taken for training. We also present a comparative study of quantile regression, differential evolution trained wavelet neural networks (DEWNN) and quantile regression random forest ensemble models in prediction in regression problems. Intervals of the parameter values of random forest for which the performance figures of the Quantile Regression Random Forest (QRFF) are statistically stable are also identified. The effectiveness of the QRFF over Quantile Regression and DWENN is evaluated on Auto MPG dataset, Body fat dataset, Boston Housing dataset, Forest Fires dataset, Pollution dataset, by using 10-fold cross validation.

Keywords: Differential Evolution trained Wavelet Neural Network, Regression, Quantile Regression, Random Forest, Quantile Regression Random Trees, Support Vector Machine.

1 Introduction

Quantification of the relationship between a response variable and a set of predictor variables is a standard problem in statistical analysis. If Y is a real-valued response variable and X is a set of independent variables based on a sample from a particular population then standard regression analysis tries to come up with an estimate $\hat{\mu}(x)$ of the conditional mean $E(Y|X = x)$ of the response variable Y, given $X = x$. The conditional mean minimizes the expected squared error loss,

$$E(Y|X = x) = arg \ \min_z E\{(Y - z)^2|X = x\},$$

Mosteller and Turkey [18] suggested that confinement to (conditional) mean of the regressand gives an incomplete picture and thus can lead to possibly

---

* Corresponding author.

M.N. Murty et al. (Eds.): MIWAI 2014, LNAI 8875, pp. 149–160, 2014.

wrong conclusions as and when assumptions of the classical linear regression model fail to hold [9]. Koenker and Basset [5] introduced a new method labelled "quantile regression" that allows the estimation of the *complete* distribution of the response variable conditional on any set of (linear) regressors.

Quantile Regression provides a framework for modelling the relationship between an outcome variable and covariates using conditional quantile functions. Quantile regression methods are based on minimizing asymmetrically weighted absolute residuals and are intended to estimate conditional quantile functions. Regression quantiles are robust against the influence of outliers.

Decision trees are widely used in data mining. Trees have the ability to handle different types of predictors, and ability to handle missing data. Fitting the tree to data is typically performed using recursive partitioning algorithms, which can be efficient and scalable [6]. Tree methods in modern use go back to the works of Breiman et al.[2] and Quinlan [3].

Random forests grow an ensemble of trees. A large number of trees are grown. For each tree and each node, random forests employs randomness when selecting a variable to split on. For regression, the prediction of random forests for a new data point is the averaged response of all trees. For details see [2].

In this paper we present the estimation methods based on Support vector trained quantile regression random forests (SVQRRF). We also present a comparative study of our proposed method with previous methods in literature and improve upon the results of Quantile Regression based predictive models, DWENN based prediction models and QRRF based prediction methods. We achieve dual goals with our proposed method viz, reducing the training data by removing the outliers from the training set and prediciting response variable with increased accuracy.

The remainder of the paper is as follows: Section 2 is devoted to literature review which discusses the origin and applications of Quantile Regression Random Forests. It also briefly presents development of DWENN and KGMDH with which we compare results of our proposed paper. In Section 3, we present a brief overview of the intelligent methods i.e., DEWNN, QR, QRRF, SVM employed in the paper. In Section 4, the experiment methodology and dataset description is provided. In Section 5 results and discussions are presented. Finally Section 6 concludes the work.

2   Literature Review

Quantile Regression was introduced by Koenker in his seminal paper [5]. After that QR has been applied in variety of fields and has been extended to other more robust models with hybridizing it already present techniques viz. Random Forest [4].

Buchinsky applied QR to study union wage effects, returns to education and labor market discrimination in U.S labor markets [28, 29]. Arias et al. [31], using data on identical twins studied interaction between observed educational attainment and unobserved ability. Eide and Showalter [32], Knight et al. [36]

and Levin [33] have studied school quality issues. Poterba and Rueben [35] and Mueller [34] studied public-private wage differentials in the United States and Canada. Apart from promising empirical studies Quantile Regression theory has also been extended in past two decades. Powell introduced the censored quantile regression model [37, 38]. This model consistently estimated conditional quantiles when observations on the dependent variable were censored. L-estimators based on a linear combination of quantile estimators have also been developed [19]. The quantile regression problem was trasformed in an optimization problem by Portnoy and Koenker [19]. Non-parametric approaches, like quantile Smoothing Splines have also been developed [20, 22]. Chaudhuri and Loh [23] developed a tree based method for estimation of conditional quantiles similar to CART [2]. Benoit and Van den Poel [47] developed a Bayesian method for binary quantile regression problem which was extended by Reddy and Ravi [44] by introducing Differential evolution trained kernel binary quantile regression (DE-KBQR) technique. Recently Meinhausen developed Quantile Regression Random Forest [4] which is based on the principles of random forest and ensembling of trees. QRRF has been tested emperically in varied domains. Francke et al. have studied estimation of suspended sediment concentration and yield [39]. Drivers of burnt areas [40], mass spectra of micro organisms [41] and uncertainty related to ramps of wind power production [42] have also been studied with QRF.

Regression problems which we analyze are based on benchmark datasets which have been previously used in literature to test many regression techniques. Farquad [43] analyzed support vector regression based hybrid rule extraction methods on Forest Fires and Boston Housing Data sets. Auto mobile, Forest Fires and Boston Housing datasets have been employed to check efficiency of Matrix-Variate Dirichlet Process Mixture Models [45]. Naveen et al. [12] have employed DWENN and Reddy and Ravi [11] have employed KGMDH on all the five data sets and got promising results. It is with these two papers that we compare our results.

3   Intelligent Techniques Overview

3.1   Support Vector Machine

A Support Vector Machine (SVM) [26] performs classification by constructing an N-dimensional hyper plane that optimally separates the data into two categories. SVM models are closely related to neural networks. Using a kernel function, SVMs are an alternative training method for polynomial, Radial Basis Function (RBF) networks and MLP classifiers, in which the weights of the network are found by solving a quadratic programming problem with linear constraints, rather than by solving a non-convex, unconstrained minimization problem, as in standard neural network training. The goal of SVM modeling is to find the optimal hyper plane that separates samples, in such a way that the samples with one category of the target variable should be on one side of the plane and the samples with the other category are on the other side of the plane. The samples near the hyper plane are the support vectors. An SVM analysis finds the hyper plane that is oriented so that the margin between the support vectors is maximized. One idea is that performance on test cases will be good if we choose the separating hyper plane that has the largest margin.

3.2 Decision Trees

Decision Trees were introduced by Quinlan et al. They are well known to be introduced here. For more details see [3].

3.3 Classification and Regression Tree (CART)

CART [2] is used to generate decision tree to solve both classification and regression problems. CART applies binary recursive split in building a tree. For the sake of brevity, details of CART are not presented here.

3.4 Quantile Regression

Quantile regression aims to estimate the conditional quantiles from data. Quantile regression was cast as an optimization problem by Koenker in [5]. Loss function is weighted with asymmetrical penalties for overfitting and underfitting. Loss function L_α is defined for $0 < \alpha < 1$ by the weighted absolute deviations.

$$L_\alpha(y,q) = \begin{cases} \alpha|y-q| & y > q \\ (1-\alpha)|y-q| & y \leq q \end{cases}$$

While the conditional mean minimizes the expected squared error loss, conditional quantiles minimize the expected loss $E(L_\alpha)$

$$Q_\alpha(x) = arg\ \min_q (\ E\ \{\ L_\alpha(Y,q)|X = x\ \}\ )$$

As descibed in [4] a parametric quantile regression is solved by optimizing the parameters so that the empirical loss is minimal. This is achieved efficiently due to the convex nature of the optimization problem [19].

In our paper we have employed Barrodale and Roberts algorithm [24] described in [25] to find conditional quantile fits which is a variation of simplex algorithm proposed by and Koenker and Barrodale.

3.5 Random Forest

An ensemble with a set of decision trees is grown in randomly selected subspaces of data and average of their prediction is used for prediction purposes. Ensembles are a divide-and-conquer approach used to improve performance. Ensemble methods work on the principal that a group of weak learners can come together to form a strong learner.

A small group of predictor variables is selected for each node to split on and the best split is calculated based on these predictor variables in the training set.

The tree is grown using CART methodology [2] to maximize size, without pruning. This randomization of subspace is combined with bagging to resample, with replacement, the training data set each time a new individual tree is grown [21].

When a new input is entered into the system, it is run down all of the trees in the ensemble grown. The prediction result may either be an average or weighted average of all of the terminal nodes that are reached. For details the reader is referred to [8].

3.6   Quantile Regression Random Forest

Quantile Regression Random Forest were introduced by Meinshausen [4]. The key difference between the random forest and quantile random forest is that in random forest for each node in each tree, it keeps only the mean of the observations that fall into this node and neglects all other information. In contrast, quantile regression forests keeps the value of all observations in this node, and assesses the conditional distribution based on this information. Algorithm proposed by Meinshausen [4] is briefly outlined here for more complete description of the algorithm and the theory reader is referred to [4].

Algorithm

1. Grow k trees $T(\theta_t)$, t = 1, . . . , k, as in random forests. For every leaf of every tree, take note of all observations in this leaf, not just their average.
2. For a given X = x, drop x down all trees. Compute the weight $w_i(x, \theta_t)$ of observation i $\in \{1, . . . , n\}$ for every tree. Compute weight $w_i(x)$ for every observation i $\in \{1, . . . , n\}$ as an average over $w_i(x, \theta_t)$, t = 1, . . . , k.
3. Compute the estimate of the distribution function for all y $\in R$, using the weights from Step 2.

Software used for this is made available by Meinshausen as a package quantregForest for R (R Development Core Team)[15]. The package builds upon the R-package randomForest [46].

3.7   DEWNN

Differential Evolution (DE) trained WNN (DEWNN) proposed by Chauhan et al. [10] consists of three layers namely input, hidden and output layers. Chauhan et al. [10] used Gaussian wavelet function viz. $f(x) = e^{-t^2}$ as an activation function in the hidden layer, while the output layer is retained as linear one. In addition to Gaussian wavelet function Chauhan et al. [10] adapted Garsons feature selection in DEWNN. We compare our work with the previous work which employed the same feature selection algorithm as Naveen et al. [12] For more details of DEWNN, the reader is referred to [10] and [12].

3.8 Group Method of Data Handling (GMDH)

The group method of data handling (GMDH) was introduced by Ivakhnenko [7] as an inductive learning algorithm for modeling of complex systems. It is a self organizing approach based on sorting out of gradually complicated models and evaluation of them using some external criterion on separate parts of the data sample [27]. The GMDH was partly inspired by research in perceptrons and learning filters. GMDH has influenced the development of several techniques for synthesizing (or self-organizing) networks of polynomial nodes. The GMDH attempts a hierarchic solution, by trying many simple models, retaining the best, and building on them iteratively, to obtain a composition (or feed-forward network) of functions as the model. The building blocks of GMDH, or polynomial nodes, usually have the quadratic form:

$$z = w_0 + w_1 x_1 + w_2 x_2 + w_3 x_1^2 + w_4 x_2^4 + w_3 x_1 x_2$$

for inputs x_1 and x_2, coefficient (or weight) vector w, and node output, z. The coefficients are found by solving the Linear Regression equations with $z = y$, the response vector.

3.9 Kernel GMDH

Ravi et al.. proposed a novel Kernel based Soft Computing hybrid viz., Kernel Group Method of Data Handling (KGMDH) to solve regression problems[11]. In the KGMDH technique, Kernel trick is employed on the input data in order to get Kernel matrix, which in turn becomes input to GMDH. In our paper we improve upon the results obtained by them on the same datasets. For details of their hybrid technique see [11].

4 Proposed Hybrid, Experiment setup and Dataset Description

4.1 Proposed Hybrid

In this paper we propose extracting support vectors from the data set and then use them to train Quantile Regression Random Forest. This serves a dual purpose viz. it reduces the data set to a more feasible size which can be handled more efficiently by QRRF and according to our experiments which are detailed below it also achieves more accurate results.

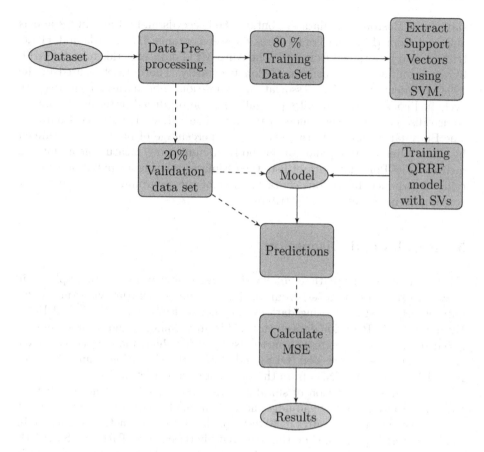

4.2   Research Methodology

All dataset are first divided into two parts of 80%: 20% ratio. After 3 runs a collection of suitable parameter values is extracted. 10 Fold Cross Validation (10 FCV) was performed on 80% part and 20% of the data to adjudge the best parameter values for a given dataset and average mean square error of 30 tests for that set of parameters is reported in Table 1.

For QRRF predictions on test data are used to tune the parameters viz. ntrees, nsize and mtry. It is noted that different values of node size do not seem to change the results very much; nevertheless, it is pointed in Lin and Jeon [13] that growing trees until each node is pure (as suggested by Breiman [2]) might lead to overfitting [4].

CRAN environment and CRAN packages "quantreg" and "quantregForest" were used to perform the numerical computation and predictions [15–17].

4.3   Data Set Description

The Boston housing dataset is taken from Statlib library which is maintained at Carnegie Mellon University. The dataset describes the housing values in the suburbs of Boston. The dataset contains 506 records and 13 attributes and is obtained from [14]. The Forest Fires dataset contains 517 records and 11 attributes and is obtained from [30]. The Auto MPG dataset is taken from Statlib library maintained at Carnegie Mellon University. The dataset is used in the 1993 American Statistical Association Exposition. The dataset concerns city cycle fuel consumption in miles per gallon, to be produced in terms of 3 multi-valued discrete and 5 continuous attributes. The dataset is obtained from [14]. The Body fat dataset lists the estimates of percentage of body fat determined by underwater weighing and various body circumference measurements for 252 men[14]. The Pollution dataset lists the estimates relating air pollution to mortality determined by various characteristics of the environment and people. It contains 60 instances and 16 attributes.

5   Results and Discussion

The effectiveness of QRRF is analyzed on regression problems by applying it to various standard data sets from Machine Learning literature viz. Auto MPG dataset [14], Boston Housing dataset [14], Forest Fires dataset [14] and Body Fat dataset [14], Pollution dataset (http://lib.stat.cmu.edu) and results are compared to three other regression methods: Quantile Regression (QR), Ordinary Least square linear regression (OLS) and Differential Evolution trained Wavelet Neural Network (DWENN) from the works of Chauhan et al.[10].

After 3 runs a collection of suitable parameter values is extracted. 10 Fold Cross verification is done to adjudge the best parameter values for a given dataset and average mean square error of 30 tests for that set of parameters is reported in Table 1. t-test is performed on the sensitivity between the QRRF and SVQRRF (t-test(1)) and also between QR and QRRF (t-test(2)) and t-statistic value is tabulated in Table 1. From the t-statistic values, it is observed that there is no statistically significant difference between the SVQRRF and QRRF at 1% level of significance but QR and QRRF differ statistically at 1% level of significance.

Table 1. Average MSE values of 10FCV of datasets

Dataset	OLS	QR	QRRF	SVQRRF	t-test(1)	t-test(2)
Automobile	0.092	0.011	0.0084	**0.003473**	1.83	4.87
Bodyfat	0.016	0.007	**0.0009**	0.001583	0.04	5.71
Pollution	0.052	**0.033**	0.0125	0.001211	1.67	2.65
Forest Fires	0.012	0.013	0.015	**0.00045**	2.7	8.11
Boston Housing	0.024	0.008	0.0070	**0.00382**	1.57	3.12

From Table 1, it is observed that SVQRRF outperforms QRRF and others on all but on dataset. But t-test conducted at 1% level of significance and 18 degrees of freedom between values of QRRF and SVQRRF leads to the conclusion that these two methods are not statistically different. It is also observed that QRRF outperforms QR in all datasets except Forest Fires and also that QR and QRRF are statistically different. It can be concluded that QRRF is a better prediction method than QR and OLS. In Table 2 CPU time taken by both the methods is listed and includes time required for extraction of support vectors in SVQRRF.

Table 2. CPU time on 2.3GHz AMD Quad-Core A10-4600M Accelerated Processor in seconds

Dataset	QRRF	SVQRRF
Automobile	0.15197	**0.13487**
Bodyfat	0.08653	**0.07921**
Pollution	**0.00695**	0.00779
Forest Fires	0.14365	**0.12955**
Boston Housing	0.16499	**0.14437**

From Table 2, it is observed that SVQRRF outperforms QRRF on four datasets out of the five datasets used. SVQRRF does this by excluding outliers from the trianing dataset and employing only the support vectors. We can also conclude that extracting support vectors and trainging QRRF on SVs is less resource consuming than trianing QRRF on the whole dataset.

For comparison with KGMDH the MSE values are reported from [11] which implemented Kernel method using Java (JDK 1.5) on Windows 7 platform on

Table 3. Average MSE values of 10FCV

Dataset	DWENN	KGMDH	SVQRRF
Automobile	0.011	0.0044	**0.003473**
Bodyfat	0.005	0.0038	**0.001583**
Pollution	0.0129	0.0074	**0.001211**
Forest Fires	0.009	0.0027	**0.00045**
Boston Housing	0.0179	0.0044	**0.00382**

machine with a RAM of 2GB. Reddy and Ravi [11] employed GMDH using the tool NeuroShell 2.

In SVQRRF on tuning the variables with help from predictions on test data, it was found that ntree = 100 to 400 resulted in better performance. Node size was kept 10. We employed Radial Basis kernel function (RBF) to extract support vectors.

From Table 3, it is observed that SVQRRF outperforms DWENN and KG-MDH in all datasets in terms of MSE. In all datasets MSE values show considerable reduction with SVQRRF than DWENN and KGMDH.

6   Conclusion

A novel hybrid using SVM and QRRF in tandem for solving regression problems is proposed. We observed that the proposed architecture SVQRRF outperformed the KGMDH, DWENN, QRRF and QR in terms of MSE for all the regression datasets. The t-test conducted between SVQRRF and QRRF, also suggested that there is no statistically significant difference between them, in majority of the cases. Hence, we can conclude that proposed support vector technique SVQRRF is a viable option for building model for regression datasets which increases the accuracy by accounting for outliers without disturbing the inherent statistical distribution of Quantile Regression Forests. Since it was seen that QRRF and SVQRRF were not statistically different, we performed CPU time testing for both the methods and it was observed that SVQRRF outperformed QRRF in time taken for the method for prediction.

References

1. Koenker, R.: Quantile Regression. Cambridge University Press (2005)
2. Breiman, L., Friedman, J., Stone, C.J., Olshen, R.A.: Classification and Regression Trees. Chapman & Hall/CRC (1984)
3. Quinlan, J.R.: C4.5: Programs for Machine Learning. Morgan Kaufmann (1993)
4. Meinhausen, N.: Quantile Regression Forest. Journal of Machine Learning Research, 983–999 (2006)
5. Koenker, R., Basset, G.: Regression Quantiles. Econometrica: Journal of the Econometric Society Econometrica, 33–50 (1978)
6. Bhat, H.S., Kumar, N., Vaz, G.: Quantile Decision Trees (2011), http://faculty.ucmerced.edu/hbhat/BhatKumarVaz2011.pdf
7. Ivakhnenko, A.G.: The group method of data handling - A rival of the method of stochastic approximation. Soviet Automatic Control 13(3), 43–55 (1966)
8. Breiman, L.: Random forests. Machine Learning 45, 5–32 (2001)
9. Schulze, N.: Applied Quantile Regression: Microeconometric, Financial, and Environmental Analyses. PhD Dissertation, Faculty of Economics, Eberhard Karls University of Tübingen (2004)
10. Chauhan, N., Ravi, V., Karthikchandra, D.: Differential evolution trained wavelet neural networks: Application to bankruptcy prediction in banks. Expert Systems with Application 36, 7659–7665 (2009)

11. Reddy, K.N., Ravi, V.: Kernel Group Method of Data Handling: Application to Regression Problems. In: Panigrahi, B.K., Das, S., Suganthan, P.N., Nanda, P.K. (eds.) SEMCCO 2012. LNCS, vol. 7677, pp. 74–81. Springer, Heidelberg (2012)

12. Naveen, N., Ravi, V., Rao, C.R.: Rule extraction from DEWNN to Solve Classification and Regression Problems. In: Panigrahi, B.K., Das, S., Suganthan, P.N., Nanda, P.K. (eds.) SEMCCO 2012. LNCS, vol. 7677, pp. 206–214. Springer, Heidelberg (2012)

13. Lin, Y., Jeon, Y.: Random Forest and adaptive nearest neighbors, Technical Report 1055, University of Wisconsin (2002)

14. Bache, K., Lichman, M.: UCI Machine Learning Repository. University of California, School of Information and Computer Science, Irvine, CA, http://archive.ics.uci.edu/ml

15. R Core Team. R: A language and environment for statistical computing. R Foundation for Statistical Computing, Vienna, Austria (2013), http://www.R-project.org/

16. Koenker, R.: quantreg: Quantile Regression. R package version 5.05 (2013), http://CRAN.R-project.org/package=quantreg

17. Meinhausen, N.: quantregForest: Quantile Regression Forests. R package version 0.2-3 (2012), http://CRAN.R-project.org/package=quantregForest

18. Mosteller, F., Tukey, J.W.: Data Analysis and Regression: A Second Course in Statistics. Addison-Wesley (1977)

19. Portnoy, S., Koenker, R.: The gaussian hare and the laplacian tortoise: Computability of squared-error versus absolute-error estimates. Statistical Science 12, 279–300 (1997)

20. He, X., Ng, P., Portnoy, S.: Bivariate quantile smoothing splines. Journal of the Royal Statistical Society B 3, 537–550 (1998)

21. Biau, G.: Analysis of a Random Forests Model. Journal of Machine Learning Research 13, 1063–1095 (2012)

22. Koenker, R., Ng, P., Portnoy, S.: Quantile smoothing splines. Biometrika 81, 673–680 (1994)

23. Chaudhuri, P., Loh, W.: Nonparametric estimation of conditional quantiles using quantile regression trees. Bernoulli 8, 561–576 (2002)

24. Barrodale, I., Roberts, F.: Solution of an overdetermined system of equations in the 1 norm. Communications of the ACM, 17, 319–320

25. Koenker, R., d'Orey, V.: Computing Regression Quantiles. Applied Statistics 36, 383–393 (1987)

26. Cortes, C., Vapnik, V.: Support-vector networks. Machine Learning (1995)

27. Srinivasan, D.: Energy demand prediction using GMDH networks. Neurocomputing 72(1-3), 625–629 (2008)

28. Buchinsky, M.: Changes in U.S. Wage Structure 1963–1987: An Application of Quantile Regression. Econometrica 62(2), 405–458 (1994)

29. Buchinsky, M.: The Dynamics of Changes in the Female Wage Distribution in the USA: A Quantile Regression Approach. Journal of Applied Econometrics 13(1), 1–30 (1997)

30. Cortez, P., Morais, A.: A Data Mining Approach to Predict Forest Fires using Meteorological Data. In: Neves, J., Santos, M.F., Machado, J. (eds.) New Trends in Artificial Intelligence, Proceedings of the 13th EPIA 2007 - Portuguese Conference on Artificial Intelligence, Guimaraes, Portugal, pp. 512–523. APPIA (2007), http://www.dsi.uminho.pt/~pcortez/fires.pdf, ISBN-13 978-989-95618-0-9

31. Arias, O., Hallock, K.F., Sosa-Escudero, W.: Individual Heterogeneity in the Returns to Schooling: Instrumental Variables Quantile Regression Using Twins Data. Empirical Economics 26(1), 7–40 (2001)
32. Eide, E., Showalter, M.: The Effect of School Quality on Student Performance: A Quantile Regression Approach. Economics Letters 58(3), 345–350 (1998)
33. Levin, J.: For Whom the Reductions Count: A Quantile Regression Analysis of Class Size on Scholastic Achievement. Empirical Economics 26(1), 221–246 (2001)
34. Mueller, R.: Public- and Private-Sector Wage Differentials in Canada Revisited. Industrial Relations 39(3), 375–400 (2000)
35. Poterba, J., Rueben, K.: The Distribution of Public Sector Wage Premia: New Evidence Using Quantile Regression Methods, NBER Working Paper No. 4734 (1995)
36. Knight, K., Bassett, G., Tam, M.S.: Comparing Quantile Estimators for the Linear Model (2000) (preprint)
37. Newey, W., Powell, J.: Efficient Estimation of Linear and Type I Censored Regression Models Under Conditional Quantile Restrictions. Econometric Theory 6, 295–317 (1990)
38. Newey, W., Powell, J.: Asymmetric Least Squares Estimation and Testing. Econometrica 55, 819–847
39. Francke, T., López-Tarazón, J.A., Schröder, B.: Estimation of suspended sediment concentration and yield using linear models, random forests and quantile regression forests. Hydrological Processes
40. Archibald, S., Roy, D.P., Wilgen, V., Brian, W., Scholes, R.J.: What limits fire? An examination of drivers of burnt area in Southern Africa. Global Change Biology 15(3), 613–630 (2009)
41. Satten, G.A., Datta, S., Moura, H., Woolfitt, A.R., Carvalho, M.G., Carlone, G.M., De, B.K., Pavlopoulos, A., Barr, J.R.: Standardization and denoising algorithms for mass spectra to classify whole-organism bacterial specimens. Bioinformatics 20(17), 3128–3136 (2004)
42. Bossavy, A., Girard, R., Kariniotakis, G.: Forecasting uncertainty related to ramps of wind power production. In: European Wind Energy Conference and Exhibition, EWEC 2010, vol. 2 (2010)
43. Farquad, M.A.H., et al.: Support vector regression based hybrid rule extraction methods for forecasting. Expert Systems with Applications 37(8), 5577–5589 (2010)
44. Reddy, K.N., Ravi, V.: Differential evolution trained kernel principal component WNN and kernel binary quantile regression: Application to banking. Knowledge-Based Systems 39, 45–56 (2012)
45. Zhang, Z., Dai, G., Jordan, M.I.: Matrix-variate Dirichlet process mixture models. In: International Conference on Artificial Intelligence and Statistics, pp. 980–987 (2010)
46. Liaw, A., Wiener, M.: Classification and Regression by randomForest. R News 2(3), 18–22 (2002)
47. Benoit, D.F., Van den Poel, D.: Binary quantile regression: A Bayesian approach based on the asymmetric Laplace distribution. J. Appl. Econ. 27, 1174–1188 (2012)

Clustering Web Services on Frequent Output Parameters for I/O Based Service Search

Lakshmi H.N. and Hrushikesha Mohanty

University of Hyderabad and CVR College of Engineering,
Hyderabad, India
hnlakshmi@gmail.com, hmcs_hcu@yahoo.com

Abstract. As growing number of services are being available, selecting the most relevant web service fulfilling the needs of the consumer is indeed challenging. Often consumers may be unaware of precise keywords to retrieve the required services satisfactorily and may be looking for services capable of providing certain outputs. We propose an approach for clustering web services in a registry by utilizing *Frequent Service Output Parameter Patterns*. The approach looks promising since it provides a natural way of reducing the candidate services when we are looking for a web service with desired *output pattern*. The experimental results demonstrate the performance of our clustering approach in our Extended Service Registry [5] and the variety of user queries supported.

Keywords: Service Registries, Service Search, Clustering web services, Output Parameter Patterns.

1 Introduction

Web services are self-contained, self-describing, modular applications that can be published, located, and invoked across the Web. As growing number of services are being available, selecting the most relevant web service fulfilling the requirements of a user query is indeed challenging. Various approaches can be used for service search, such as, searching in UDDI, Web and Service portals. In this paper we take up the issue of searching at service registries for its practicality in business world as providers would like to post their services centrally, as searching there is less time consuming than searching on world wide web.

Often consumers may be unaware of exact service names that's fixed by service providers. Rather consumers being well aware of their requirements would like to search a service based on their commitments(inputs) and expectations (outputs). Based on this concept we have explored the feasilbility of I/O based web service search in our previous work, where we propose an Extended Service Registry [5], to support varying requirements of the consumer. Utility of such an I/O based web service search for composition of webservices is also shown in our previous work[5]. Further, to accelerate parameter based search in our Extended Service Registry, in this paper we propose an approach to cluster services based on their output parameter pattern similarity.

M.N. Murty et al. (Eds.): MIWAI 2014, LNAI 8875, pp. 161–171, 2014.

Feature based clustering of services in a web registry reduces search space while looking for a service from the registry to meet a search requirement. The methods proposed in [3,4,6,9,11] for service clustering uses standard clustering algorithms like agglomerative clustering[9,11], Quality Threshold clustering[4] and Optics algorithm.

The current approaches employ clustering during the service search process, to cluster search results for a given user query. Mostly, the approach[3] of query processing follows two basic steps - reducing the domain of search by service domain e.g. hotel, transport etc. and then grouping the services based on query requirements. As these are done during processing of a query, it becomes expensive in terms of time. Instead, we propose an approach that does a book keeping of services following an order so that the pre-processing helps to reduce search time.

Our clustering approach exploits the co-occurrence of output parameters to cluster web services. Set of output parameters that co-occur in more than a threshold number of web services make a *Frequent Service Output Parameter Pattern*. From the many *Frequent Service Output Parameter Patterns* generated we need to choose a subset such that cover of the subset consists of all web services in the registry and also the clusters generated have a minimum overlap. We define Cluster Overlap of candidate clusters based on the number of *Frequent Service Output Parameter Patterns* that the services in the candidate cluster is supporting. Further, a preference factor is used to suggest the inclusion of clusters into final clustering.

The rest of the paper is organized as follows. In Section 2 we describe our approach for clustering services on *Frequent Service Output Parameter Pattern*. Section 3 discusses our experimental results. In Section 4 we essay the related work. We conclude our work in Section 5.

2   Clustering Services on Frequent Service Output Parameter Patterns

In data mining, association rule mining is a popular and well researched method intended to extract interesting correlations, frequent patterns, associations among sets of items in large databases or other data repositories. Frequent itemsets form the basis of association rule mining. Apriori[1] is a classic algorithm for frequent itemset mining and association rule learning over transactional databases.

A clustering approach for web services based on *frequent service output parameter patterns* looks promising since it provides a natural way of reducing the candidate services when we are looking for a web service with desired *output parameter pattern*. Since we are dealing with web services, we will use the notion of *output parameter sets*. We use the term *service output parameter pattern* to represent a *output parameter set*. The key idea is to consider the *frequent service output parameter patterns* as cluster candidates and produce a flat clustering covering all web services in the registry.

2.1 Definitions

Let $R = \langle P, W \rangle$ be a service registry , where P is a set of parameters , $P = \{p_1, p_2, .., p_n\}$ and W is a set of web services in the registry, $W = \{ws_1, ws_2, ..., ws_n\}$. Each web service, ws_i in W, has typically two sets of parameters as set of inputs ws_i^I and set of outputs ws_i^O, from P.

- **Service Output Parameter Pattern:** We use the term *service output parameter pattern* to represent a set of output parameters, i.e, $O_P \subseteq P$.

- **Cover:** For any service output parameter pattern, $O_P \subseteq P$, let $C(O_P)$ denote the cover of O_P, defined as the set of all services supporting O_P, i.e, $C(O_P) = \{ws_i \in W | O_P \subseteq ws_i^O\}$.

- **MinSupp:** Let $MinSupp$ be a real number, such that $0 \leq MinSupp \leq 1$, representing the minimum support used in the mining process to generate the set of all *frequent service output parameter patterns*. The value of *MinSupp* governs the minimum number of services that is to be present in a candidate cluster.

- **Frequent Service Output Parameter Pattern:** Let $S = \{S_1, S_2, .., S_k\}$ be the set of all *frequent service output parameter patterns* in W with respect to $MinSupp$.

- **Candidate Cluster:** The cover of each $S_j \in S$, i.e , $C(S_j)$ can be regarded as a *Candidate Cluster*.

- **Service Cluster Description:** We define *Service Cluster Description* , SCD , as a subset of S that minimally covers the entire registry W, i.e,

$$SCD = \{S_j | S_j \in S\} \text{ such that } \bigcup C(S_j) = W$$

A set of services that cover a frequent service output parameter pattern is a *candidate Cluster*. To choose an appropriate subset from the set of all *frequent service output parameter patterns* we propose to use the mutual overlap between services supporting the different *frequent service output parameter patterns*.

Our approach assumes that each web service in the registry supports at least one *frequent service output parameter pattern*. We need to determine a *Service Cluster Description* with a minimum overlap of the clusters. To measure the overlap between the different *Candidate Clusters* we propose to use the mutual overlap between the services supporting different *Candidate Clusters*. Further, a preference factor is used to suggest the inclusion of a *Candidate Cluster* into *Service Cluster Description*.

Our approach uses a greedy algorithm which initially selects a *Candidate Cluster* that has the least cluster overlap, from the set of other *Candidate Clusters*, and includes them in SCD if its contribution to SCD is greater than the preference factor. The process is repeated until SCD covers the entire registry

W. So for a given registry R involving P parameters, SCD covers all possible queries on *service output parameter patterns*.

Let NFS_i denote the number of *frequent output parameter patterns* supported by a web service ws_i ,i.e,

$$NFS_i = \{|S_j \in S|S_j \subseteq ws_i^O\}, \forall ws_i \in W.$$

Cluster Overlap: We define *Cluster Overlap* of a *Candidate Cluster* C_j, representing a *frequent output parameter pattern* S_j, denoted by $CO(C_j)$, as the average $NFS_i - 1$ value of all web services in the cluster,i.e,

$$CO(C_j) = \frac{\sum_{ws_i \in C_j} (NFS_i - 1)}{|C_j|} \qquad (1)$$

Smaller the values of NFS_i are, smaller will be the value of *cluster overlap* of a cluster C_j. Ideally, if $NFS_i = 1$ for all services in the cluster C_j then C_j has an overlap of 0 with the other candidate clusters.

To minimize the effect of monotonicity property of *frequent output parameter pattern* sets (each subset of a *frequent output parameter pattern* set is also frequent), we have defined the *cluster overlap* of clusters based on the *frequent output parameter pattern* that the clusters represent. Also, we include only those clusters into *Service Cluster Description* that results in a sufficient change in cover of SCD, thereby ignoring clusters that have services already included in the cover.

Preference Factor: Let $|SCD|$ represent the number of services, excluding repetitions, in the clusters currently present in *Service Cluster Description, SCD*. After calculating *cluster overlap* for all the cluster candidates , we choose a cluster candidate having the least overlap value, C_j , and add it to SCD if inclusion of C_j into SCD will increment the size of *Service Cluster Description* by a factor greater than or equal to *preference factor*(pf) i.e, we include the cluster C_j into SCD if it satisfies the below relation,

$$\frac{(|SCD \cup C_j| - |SCD|)}{(|W|)} \geq pf, \ 0 \leq pf \leq 1 \qquad (2)$$

The process is repeated for all *frequent output parameter patterns*, until the *Service Cluster Description,SCD*, covers all the web services in the registry, i.e, $|SCD| = |W|$.

2.2 Frequent Service Output Parameter Pattern Based Clustering Algorithm

As defined earlier *Service Cluster Description, SCD*, is a subset of the set of all *frequent service output parameter patterns*, S, that covers the entire service registry W. Due to the inherent complexity of the *frequent output parameter pattern* based clustering approach: the number of all subsets of S is $O(2^{|S|})$ an exhaustive search is prohibitive, hence we present a greedy approach for discovering *Service Cluster Description*.

Algorithm 1. Frequent Service Output Parameter Pattern based Clustering

---

Input: W, MinSupp, f
Output: Service Cluster Description,SCD
1 SCD = {}
   // Find all Frequent Service Output Parameter Patterns
2 S= DetermineS(W,MinSupp)
3 **forall the** $S_i \in S_k$ **do**
4   Calculate $CO(S_i)$ using **eqn 1**
5 **forall the** $S_i \in S$ **do**
   // Let CC represent Chosen Cluster
6   CC = S_i with least CO
7   **if** CC *satisfies* **eqn 2 then**
8   SCD = SCD $\cup$ CC
9          $S = S$ - S_i
10   **if** $(|SCD| = |W|)$ **then**
11   break
12 **if** $(|SCD| \neq |W|)$ **then**
13   **while** $(|SCD| \neq |W|)$ **do**
14   CC = S_i with least CO
15   SCD = SCD $\cup$ CC
16         $S = S$ - S_i
17 **return** $[SCD]$

Algorithm $FSOPC$, *Frequent Service Output Parameter Pattern based Clustering*, determines a *Service Cluster Description* that covers the entire service registry W with minimum *cluster overlap*. Algorithm 1 presents a pseudo-code for *Frequent Service Output Parameter Pattern based Clustering*.

Table 1 illustrates the frequent patterns obtained on a sample set of web services containing 12 web services. The highlighted rows indicates the working of Algorithm $FSOPC$, *Frequent Service Output Parameter Pattern based Clustering* on the sample set with *MinSupp=0.16* and $pf = 0.1$.

2.3   Accelerating Parameter Based Service Search

The clusters obtained from Algorithm $FSOPC$ are stored in a ClusterTable to accelerate parameter based service search. Algorithm 2 presents pseudo-code for Output Parameter based Service search utilizing the ClusterTable. Q^O represent output parameters specified in user query.

Table 1. Illustration of Algorithm *FSOPC* for Sample Web Services

Service Output Parameter Pattern	Cluster Candidates	Cluster Overlap
Period	ws6,ws8	0.5
HotelName	ws1,ws10,ws11	9.33
FlightInfo	ws2,ws10,ws11	9.33
CarType	ws3,ws10,ws11	9.0
TourInfo	ws4,ws8,ws12	0.66
PackageID	ws9,ws12	0.5
TaxiCost	ws5,ws11	8.0
TourCost	ws7,ws10,ws11	8.66
HotelCost	ws1,ws11	9.0
FlightCost	ws2,ws11	9.0
HotelName,HotelCost	ws1,ws11	9.0
FlightInfo,FlightCost	ws2,ws11	9.0
CarType,TaxiCost	ws3,ws11	8.5
HotelName,FlightInfo	ws10,ws11	13
HotelName,CarType	ws10,ws11	13
HotelName,TourCost	ws10,ws11	13
HotelName,FlightInfo,CarType	ws10,ws11	13
HotelName,FlightInfo,TourCost	ws10,ws11	13
FLightInfo,CarType,TourCost	ws10,ws11	13
HotelName,FlightInfo,CarType,TourCost	ws10,ws11	13

3   Experimental Results

The effectiveness of Frequent Service Output Parameter Pattern based Clustering method is shown by conducting two sets of experiments:

1. Performance Analysis of our clustering approach.
2. Performance improvement obtained using clustering for Output Parameter based service search.

Experimental Setup: We conducted experiments on QWS-WSDL Dataset[7]. We ran our experiments on a 1.3GHz Intel machine with 4 GB memory running Microsoft Windows 7. Our algorithms were implemented using Oracle 10g and JDK 1.6. Each query was run 5 times and the results obtained were averaged, to make the experimental results more sound and reliable.

3.1   Performance Analysis

In this section, we analyze the performance of Algorithm *FSOPC* in terms of Number of clusters in the final *Service Cluster Description,SCD*, and the average Cluster Overlap in *SCD*. The number of web services were varied from 100 to 500 and a constant value of 0.01 was used for *MinSupp*.

Algorithm 2. Service Search Using Clusters

Input: Q^O, ClusterTable : table
Output: MWSTable : table
1 **foreach** *ParName in Q^O* **do**
2   Select PID from ParTable where PName = *ParName*
3   INSERT PID into the QueryT table
4 **foreach** *Cl in ClusterTable* **do**
5   cntWS = Number of output parameters in *Cl.ClusterPars* matching those in Q^O
6   **if** *cntWS ¿ 0* **then**
7   **foreach** *ws in Cl.WSList* **do**
8   **if** *$ws^O = Q^O$* **then**
9   INSERT *ws* as Exact Match in MWSTable
10   **else if** *$ws^O \subset Q^O$* **then**
11   INSERT *ws* as Partial Match in MWSTable
12   **else**
13   INSERT *ws* as Super Match in MWSTable

Impact of Preference factor on Number of Clusters. In first, we compared the number of clusters obtained in the final *SCD* with respect to the value of *preference factor*. The results for Algorithm *FSOPC* for 100 web services and 500 web services are shown in Figure 1.

Impact of Preference factor on Cluster Overlap. Next, we compared the average Cluster Overlap in final *SCD* with respect to the value of *preference factor*. The results are shown in Figure 2. It is seen that average Cluster Overlap in both the cases (100 web services and 500 web services) decreases as the *preference factor* increases upto a value of around 0.03 and increases thereon,

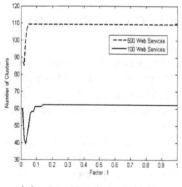

(a) pf vs Number of Clusters

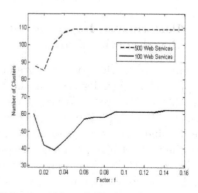

(b) pf vs No. of Clusters zoomed for f upto 0.16

Fig. 1. Preference Factor vs Number of Clusters in SCD

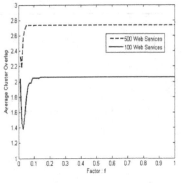

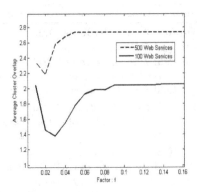

(a) pf vs Average Cluster Overlap

(b) pf vs Avg Cluster Overlap zoomed for f upto 0.16

Fig. 2. Preference Factor vs Average Cluster Overlap for clusters in SCD

as visualized in 2b. Also the value of average Cluster Overlap remains constant after the value of *preference factor* crosses 0.05 for 100 web services and 0.09 for 500 web services, as seen in 2a.

From the results obtained, we can infer that the *preference factor* is indeed required to have a control on the number of clusters obtained by Algorithm *FSOPC*, but as the value of *preference factor* crosses a cut-off point it will have the same effect on number of clusters created. Also the value of *preference factor* affects the *Cluster Overlap* among the clusters generated by the algorithm. The cut-off point for *preference factor* depends on the dataset and the number of web services. From the results obtained and the analysis made above we come to a conclusion that a value between 0.01 and 0.03 is ideal for the *preference factor* and using Algorithm *FSOPC* with this ideal value we will obtain minimum number of clusters in *SCD* with minimum average *Cluster Overlap*.

3.2 Performance Improvement Obtained in Output Parameter Based Service Search

To improve the performance of parameter based service search, we have proposed an approach to cluster services in the registry on their output parameter patterns as discussed in section 2. We conducted experiments to evaluate the performance of Output Parameter Based Service Search and also to evaluate the performance gain obtained with the inclusion of clustering technique.

In the first set of experiments we published 500 web services in our Extended Service Registry[5]. We then evaluated the performance of parameter based search using different output parameter patterns as user queries. To evaluate the performance of Output Parameter based Service Search Using Clusters, we first implemented Algorithm 1 to obtain clusters of the registered services and then implemented Algorithm 2 on these clusters. We then queried the registry with the same set of output parameter patterns as before. The experiment was repeated by publishing 1000 web services in Service Registry. Figure 3 shows that on integrating our clustering approach into the service registry, there is a substantial improvement in the performance of Parameter based Service Search.

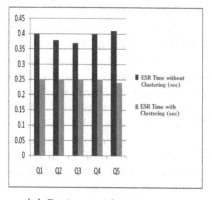

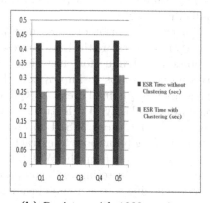

(a) Registry with 500 services   (b) Registry with 1000 services

Fig. 3. Output Parameter based Search Using Clusters

4   Related Work

Specifications for web services fall into two main categories viz. ontology based and keyword based. A web service of former category is built with domain ontology whereas of latter case keywords are used for specifying inputs and outputs. Though, semantic web services are versatile in arture still keyword based service specification is popular for its fast response avoiding reference to domain ontology that could be at times large and complex enough for the purpose. Hence we have taken keyword based service specification and are interested in pattern search that correspond to service functionality which user may find interesting.

Research in clustering web services has recently gained much attention due to the popularity of web services and the potential benefits that can be achieved from clustering web services like reducing the search space of a service search task. There are a number of approaches proposed in recent years for non-semantic web service search[4,6,9,11,12]. In this section we summarize various clustering approaches for non-semantic web services.

Xin Dong et.al.[11] develop a clustering based web service search engine, Woogle. Their search approach consists of two main phases: first, the search engine returns a list of web services that match a given user query, specified as a set of keywords. Second, their tool extracts a set of semantic concepts, by exploiting the co-occurrence of terms in web service inputs and outputs of operations, to cluster terms into meaningful concepts by applying an agglomerative clustering algorithm. This makes it possible to combine the original keywords with the extracted concepts and compare two services on a keyword and concepts level to improve precision and recall. This approach leads to significantly better results than a plain keyword-based search. Richi Nayak and Bryan Lee[9] attempts to handle the problem by suggesting to the current user with other related search terms based on what other users had used in similar queries by applying clustering techniques. Web services are clustered based on search sessions instead of individual queries.

For the calculation of non-semantic similarity between web services, WSDL-based approaches are the most representative work[4,6]. Khalid Elgazzar et.al.[4] propose a technique for clustering WSDL documents into functionally similar web service groups to improvize web service search in service search engines. They use the Quality Threshold (QT) clustering algorithm to cluster web services based on similarity of 5 features in WSDL documents which include - WSDL content, WSDL types, WSDL messages, WSDL ports and the web service name. Liu and Wong[6] propose to extract 4 features, i.e., content, context, host name, and service name, from the WSDL document to cluster web services. They take the process of clustering as the preprocessor to search, hoping to help in building a search engine to crawl and cluster non-semantic web services.

The approaches discussed above employ clustering during the service search process, to cluster search results for a given user query. Mostly, the approach [3] of query processing follows two basic steps - reducing the domain of search by service domain e.g. hotel, transport etc. and then grouping the services based on query requirements. As these are done during processing of a query, it becomes expensive in terms of time. Instead, we propose an approach that does a book keeping of services following an order so that the pre-processing helps to reduce search time. The order here is formed as patterns of outputs i.e co-occurence of output parameters. Though, time is spent in pre-processing still, the gain in the proposed scheme is better than the time due to the approach followed earlier[3]. Our approach is useful when we are looking for a web service with desired output pattern.

5   Conclusion

Query based web service discovery is an important issue, especially for non-semantic web services. Traditional UDDI based service search lacks the ability to recognize the features described in WSDL files. In this paper, we propose an approach to improve service search of non-semantic web services by clustering services on their output similarity. Our approach for clustering web services based on *frequent service output patterns* looks promising since it provides a natural way of reducing the candidate services when we are looking for a web service with desired *output pattern*.

Algorithm *FSOPC* in section 2.2 generates a *Service Cluster Description* that covers all the web services in the registry. The cluster selection in the algorithm is governed by *preference factor*, as discussed in section 2.1. We have simulated the algorithms on QWS-WSDL Dataset[7]. The experimental results demonstrate the performance of our clustering approach on varying user queries.

We have proposed an approach to widen the scope of service composition reported in our previous work [5], considering three possible modes of service composability: Exact, Partial and Super. In the future work we would like to work on a strategy for choosing the best matching service among the many candidate services. We further plan to integrate the current proposal with our previous work, to generate all possible compositions for a given user query, when the query is expressed in terms of I/O parameters. Further, we plan to study the adaptability of the clustering approach as web services are dynamically added on to the registry.

References

1. Agrawal, R., Srikant, R.: Fast algorithms for mining association rules in large databases. In: Bocca, J.B., Jarke, M., Zaniolo, C. (eds.) Proceedings of the 20th International Conference on Very Large Data Bases (VLDB), Santiago, Chile, pp. 487–499 (September 1994)
2. Apache jUDDI, http://juddi.apache.org/index.html
3. Ma, J., Zhang, Y., He, J.: Efficiently finding web services using a clustering semantic approach. In: Proceedings of the 2008 International Workshop on Context Enabled Source and Service Selection, Integration and Adaptation: Organized with the 17th International World Wide Web Conference (WWW 2008), p. 5. ACM (April 2008)
4. Elgazzar, K., Hassan, A.E., Martin, P.: Clustering wsdl documents to bootstrap the discovery of web services. In: 2010 IEEE International Conference on Web Services (ICWS), pp. 147–154 (July 2010)
5. Lakshmi, H.N., Mohanty, H.: RDBMS for service repository and composition. In: 2012 Fourth International Conference on Advanced Computing (ICoAC), December 13-15 (2012)
6. Wei, L., Wong, W.: Web service clustering using text mining techniques. International Journal of Agent-Oriented Software Engineering 3(1), 6–26 (2009)
7. QWS-WSDL Dataset, http://www.uoguelph.ca/qmahmoud/qws
8. Ran, S.: A Model for Web Services Discovery With QoS. ACM Sigecom Exchanges 4(1), 1–10 (2003)
9. Nayak, R., Lee, B.: Web Service Discovery with additional Semantics and Clustering. In: 2007 IEEE/WIC/ACM International Conference on Web Intelligence (2007)
10. UDDI Specifications, http://uddi.org/pubs/uddi_v3.html
11. Dong, X., Halevy, A., Madhavan, J., Nemes, E., Zhang, J.: Similarity search for web services. In: Proceedings of the Thirtieth International Conference on Very Large Data Bases, vol. 30, pp. 372–383. VLDB Endowment (August 2004)
12. Yu, J., Liu, F., Shi, Y., Wang, T., Wu, J.: Measuring similarity of web services based on wsdl. In: Proceedings of International Conference on Web Services, pp. 155–162 (2010)

IntelliNavi: Navigation for Blind Based on Kinect and Machine Learning

Alexy Bhowmick[1], Saurabh Prakash[1], Rukmani Bhagat[1],
Vijay Prasad[1], and Shyamanta M. Hazarika[2]

[1] School of Technology, Assam Don Bosco University
Guwahati, Assam, India
{alexy.bhowmick,saurabhp,rukmanib,vijay.prasad}@dbuniversity.ac.in
[2] School of Engineering, Tezpur University
Tezpur, Assam, India
smh@tezu.ernet.in

Abstract. This paper presents a wearable navigation assistive system for the blind and the visually impaired built with off-the-shelf technology. Microsoft Kinect's on board depth sensor is used to extract Red, Green, Blue and Depth (RGB-D) data of the indoor environment. Speeded-Up Robust Features (SURF) and Bag-of-Visual-Words (BOVW) model is used to extract features and reduce generic indoor object detection into a machine learning problem. A Support Vector Machine classifier is used to classify scene objects and obstacles to issue critical real-time information to the user through an external aid (earphone) for safe navigation. We performed a user-study with blind-fold users to measure the efficiency of the overall framework.

Keywords: Kinect, RGB-D, Blind, Navigation systems, Object recognition, Machine Learning.

1 Introduction

The blind and visually impaired (VI) face many challenges in their everyday life. People with vision disabilities are handicapped in perceiving and understanding the physical reality of the environment around. Searching, walking, crossing streets, recognizing objects, places and people becomes difficult or impossible without vision. Hence support tools become absolutely essential while performing activities of daily living. Recent statistics from the World Health Organization estimate that there are 285 million people worldwide who are visually impaired: 39 million are *blind* and 246 million have *low vision*; 90% of world's visually impaired live in developing countries [3]. Herein lies the motivation for innovation of affordable tools to aid the affected community. Research in the area indicates that Computer Vision embodies a powerful tool for the development of assistive technologies for the blind[10],[18],[11]. Reliable and robust vision systems for the blind and visually impaired involving cutting-edge technology, provided at an affordable cost can have a very relevant social impact.

M.N. Murty et al. (Eds.): MIWAI 2014, LNAI 8875, pp. 172–183, 2014.

One major challenge for the visually handicapped everyday is safe navigation by detecting and avoiding objects or hazards along their walking path in an indoor or outdoor environment. This paper presents a wearable computerized navigation system that uses Microsoft Kinect[2] and state-of-the-art techniques from Computer Vision and Machine Learning to detect indoor objects or obstacles and alert the blind user through audio instructions for safe navigation in an indoor environment.

Till today, the blind and visually impaired people rely heavily on their canes, guide dogs, or an assistant for navigating in an unfamiliar environment. In case of familiar environments the blind mostly depend on their sense of orientation and memory[6]. The traditional *white cane* helps the blind user to familiarize oneself with the immediate surroundings. However, this process requires memorizing the locations of doors, exits or obstacles and can be arduous, time taking, and mentally taxing. Moreover, any change in a familiar environment's configuration demands the familiarization process to be repeated again. The advent of the very affordable Microsoft Kinect sensor opens up new possibilities of creating technology which provides a degree of situational awareness to a blind person and simplifies the daily activities. Kinect[2] is a widely deployed off-the-shelf technology that has evolved to be a very suitable sensor to detect humans and objects and navigate along a path [5][15][11][16][12]. Released in the context of video games and entertainment, it is a recent device that has been applied in several areas such as – Computer Vision, Human-Computer Interaction, Robot navigation, etc. The goal of our system is to facilitate *micro-navigation*, i.e. sensing of immediate environment for obstacle and hazards. The system is affordable, and provides features such as obstacle detection, auditory assistance and navigation instructions. These features can be extremely helpful to the blind in performing their daily tasks such as walking by a corridor, recognizing an object, detecting staircases, avoiding obstacles, etc. The proposed navigation system is meant to complement the traditional navigational tools and not necessarily replace them.

The paper is organized as the follows. Section 2 discusses various methods on visual object detection in vision-based navigation systems involving Kinect. In Sect. 3 we present the system architecture of the navigation system and the proposed framework for obstacle detection and safe indoor navigation. In Sect. 4 we report on implementation and present results of a user study with few blind-folded users. The evaluation of the system showed above 82% overall object detection accuracy in a controlled environment. Section 5 presents the conclusion and possible future works.

2   Related Work

Visual object recognition of everyday objects in real world scenes or database images has been an ongoing research problem. The release of Kinect and the wide availability of affordable RGB-D sensors (color + depth) has changed the landscape of object detection technology. We review current research work in these fields - methods for addressing vision problems and development of Kinect based navigation systems specifically developed to empower the blind and VI community. Vision-based navigation is one popular approach for providing navigation to blind and visually impaired humans or autonomous robots [15].

NAVI [18] - is a mobile navigational aid that uses the Microsoft Kinect and optical marker tracking to help visually impaired (VI) people navigate inside buildings. It provides continuous vibro-tactile feedback to the persons waist, instead of acoustic feedback as used in [5] and [13], to give an impression of the environment to the user and warn about obstacles. The system is a proof-of-concept and a lacks a user study. Mann et. al. [14] present a novel head-mounted navigational aid based on Kinect and vibro-tactile elements built onto a helmet. The Kinect draws out 3D depth information of the environment observed by the user. This depth information is converted into haptic feedback to enable blind and visually impaired users to perceive depth within a range and avoid collisions in an indoor environment. Depth information has been extensively used in a variety of navigation applications. Brock and Kristensson [6] present a system that perceives the environment in front of the user with a depth camera. The system identifies nearby structures from the depth map and uses sonification to convey obstacle information to the blind user. Khan, Moideen, Lopez, Khoo, and Zhu [13] also aimed at developing a navigation system utilizing the depth information that a Kinect sensor provides to detect humans and obstacles in real time for a blind or visually impaired user. Two obstacle avoidance approaches *one-step approach* and *direction update approach* were experimented with and a user-study was performed. However, in real time, the authors reported slow execution.

Other authors have used Kinect to create a 3D representation of the surrounding environment. Bernabei, Ganovelli, Benedetto, Dellepiane and Scopigno [5] present a low-cost system for unassisted mobility of the blind that takes as input the depth maps and data from the accelerometer produced by the Kinect device to provide an accurate 3D representation of the scene in front of the user. A framework for time-critical computation was developed to analyze the scene, classify the obstacles on ground, and provide the user with a reliable feedback. KinSpace [11] is another Kinect-based passive obstacle detection for home that monitors the open space for the presence of obstacles and notifies the resident if an obstacle is left in the open space.

Some authors have employed Kinect to provide autonomy to mobile robots, such as Sales, Correa, Osrio, and Wolf [15], who present an autonomous navigation system based on a finite state machine learned by an artificial neural network (ANN) in an indoor environment. The experiments were performed with a Pioneer P3-AT robot equipped with a Kinect sensor. The authors claimed to achieve excellent results, with high accuracy level for the ANN individually, and 100% accuracy on navigation task.

The use of machine learning (ML) presents a promising new approach to process RGB-D data stream acquired by Kinect. In recent years increasing research efforts in the machine learning community has resulted in a few significant assistive systems. Filipe, Fernandes, Fernandes, Sousa, Paredes, and Barroso [10] present an ANN-based navigation system for the blind that perceives the environment in front of the user with the help of the Kinect depth camera. The system is a partial solution but is able to detect different patterns in the scene like - no obstacles (free path), obstacle ahead (wall), and stairs (up/down). The authors in [16] and [9] have both experimented with a framework based on Support Vector Machines (SVMs) combined with local descriptors to report on the performance of the classifier in categorization of varied objects such as cups, tomatoes, horses, cars, stairs, crossings, etc. Over the past decade increasing efforts and innovations have focused on generic object detection in images or real world scenes. Visual object recognition for assistive systems and vision-based navigation are fertile fields attracting many researchers with a growing interest in developing Computer Vision algorithms and mechanisms to address actual problems.

3   Proposed System - *IntelliNavi*

In this section we describe the proposed framework for recognizing indoor object from RGB-D sensor data and issuing audio feedback to the user. Feature extraction from RGB-D images is described next and finally the classification of indoor objects encountered and the navigation mechanism is presented.

Figure 1 shows the proposed Computer Vision-based framework for the detection of common indoor objects. The Kinect sensor is used to extract the RGB-Depth information of the environment being observed by the user. First, detection and description of *interest points* from training data is performed by the Speeded Up Robust Features (SURF) local descriptors. These descriptors of image patches are clustered using the Bag-of-Visual-Words (BOVW) model [8] - a novel approach to generic visual categorization to construct feature vectors. The feature vectors are of fixed dimension and serve as input to a multi-class SVM classifier to distinguish between objects.

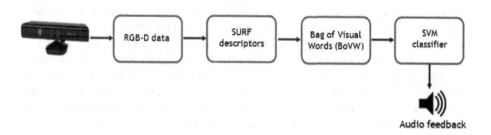

Fig. 1. Framework of the proposed navigation assistive system for blind using Kinect and Machine Learning

Thus a novel mechanism involving Kinect, RGB-D data, SURF, Bag-of-Visual-Words and SVM classifier, (which were not used together in literature before) is developed. The audio feedback program notifies the user of obstacles ahead and suggests the blind user with an instruction to move or turn to a specific route to reach the endpoint. Figure 2 shows the user description of the navigation system. The navigation system integrates a Microsoft Kinect[2], a Laptop with earphone connected, a battery pack, and a backpack construction for the laptop. The Kinect device is fixed at the height of the pelvis. The backpack construction carries the laptop as the program modules analyze the scene and process objects.

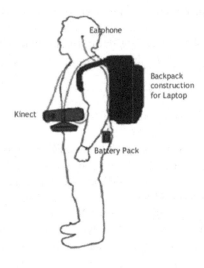

Fig. 2. User description

3.1   System Architecture

Instead of a statically placed Kinect that tracks moving objects (as in [11]), we track the static environment with a moving waist-mounted Kinect (Fig. 2). To power the mobile Kinect we use a 12V - 7Ah battery pack that provided sufficient hours of power supply during our tests. The setup may be seem to be a cumbersome computer vision system because of the weight and size of the components, but our objective was to test the technological concept, hoping that further advancements could handle the portability issues. The system has two modes: Firstly, after being trained, detect foreseen indoor objects and classify them using a state-of-the-art supervised classifier. Secondly, communicate the environment information back to the user through an audio feedback.

Data Acquisition. Kinect provides a depth sensor, an RGB camera, an accelerometer, a motor and a multi-array microphone. Kinect leads to low-cost solutions in object recognition and works well in low light as well. The RGB camera (capable of capturing data at 640x480 pixels in 30 Hz) is associated to an infrared transmitter and receiver allowing depth estimation of the environment elements. The angular ranges, covered by the depth map, are about 60 degrees (horizontal) and 40 degrees (vertical) in front of the device. We adapted the Kinect to feed its images into a PC where vision processing modules get activated by the depth sensor at approximately 1-1.5 meters distance from any object. The system determines if there are objects or obstacles in the path, how close they are, and which way should one move to avoid them.

Fig. 3. a) Depth image of a chair acquired by Kinect b) Depth image of a RGB laptop on a table. Colors indicate distance ranging from *safe* to *very close* c) SURF interest points detected in a sample RGB chair image d) SURF interest points detected in a sample RGB laptop image.

Feature Extraction from RGB-Depth Data. SURF[1] is a fast, performant scale and rotation-invariant detector and descriptor. The SURF detector provides an almost real-time computation without loss in performance making it ideal for Computer Vision applications such as ours. The SURF descriptor is based on sums of Haar wavelet components [4]. We chose to extract SURF descriptors (from the training set images) as interest points since they are robust to image distortions and transformations including rotation, scale, illumination, and changes in viewpoint. These interest points contain rich local information of the image and local image descriptors have shown very good performance in object recognition tasks [9]. SURF constructs a descriptor vector of length 64 to

---

[1] http://www.vision.ee.ethz.ch/~surf

store the local features of the RGB-D images. The SURF implementation used in this paper is provided by the Accord.NET framework [2].

Most supervised classifiers (including SVMs) require feature vectors of fixed dimensions as input [17]. To transform the raw interest points provided by SURF into an image feature with a fixed dimension we opt for the BOVW [8] representation. Bag-of-Visual-Words assigns the SURF descriptors from training images to N clusters of visual words using the K-means algorithm. This BOVW model representation of these descriptors results in fixed dimensional feature vectors perfectly suited for a supervised classifier, which is next in our framework.

Training Support Vector Machines. SVMs have faster training times and generalize very well on smaller training sets [7], hence we use SVMs to train SURF descriptors and visual words. We built a supervised classifier based on the visual word features and applied them to predict objects in the indoor scenario. In order to extend the original binary SVM to a multiple class problem like ours, we built a *one-vs-one* scheme where multiple SVMs specialize to recognize each of the available classes. A multi class problem then reduces to n*(n-1)/2 smaller binary problems. Given an object encountered in test scenario, it is then assigned to the class with the maximum votes (i.e. the class selected by the most classifiers). An alternative strategy to handle the multi class case is to train N *one-vs-rest* classifiers as described in [7],[9].

Audio Feedback - Output. Although many navigation applications for visually impaired [5][13][14], have used audio as the primary interface for providing feedback, there is a strong argument against this solution: In micro-navigation scenarios, acoustic feedback can be annoying or even distracting since we are depriving the blind person of his/her sense of hearing [5][18]. The blind and visually impaired (VI) have limited sensory channels which must not be overwhelmed by information. Hence, we tried with very brief audio instructions that are triggered by the SVM classifier. Simple audio messages (e.g. "Wait", "Obstacle ahead", "Take left", "Take right", etc) were provided via the earphone with the sole goal of leading a blind user safely from an initial point to an end-point.

4   Implementation and Experimental Results

In order to test the system, we developed a real-time application of the framework in C# using the Microsoft Kinect Software Development Kit (SDK) v1.8, Microsoft Visual Studio 2012, and Accord.NET framework v2.12. The proposed framework was tested on a Windows 8 platform running atop a 2.66 GHz processor with a 4GB RAM.

---

[2] www.accord-framework.net - a C# framework for building Machine Learning, Computer Vision, etc applications.

4.1 RGB-D Object Dataset

To experiment with the proposed framework, we developed a personal dataset consisting of images of common indoor objects e.g. chair, paper bin, laptop, table, staircase and door. We captured indoor workplace settings for our dataset, besides using subsets of the Caltech 101 dataset [1]. The training dataset is composed of 876 images with approximately 145 images representing each of the predefined objects (classes). The adopted learning algorithm is supervised, so the Support Vector Machine (SVM) was provided with labeled instances of the objects in training.

Fig. 4. Sample images of common indoor objects from our dataset

4.2 System Integration and Testing

We assembled the components mentioned in Fig. 2 for testing purposes. The Kinect was fixed at waist height with support from the neck. We performed a user-study with two blind-folded users to measure the efficiency and robustness of the algorithms and the framework. In a series of repeated experiments, the blind-folded users were instructed to walk along a path, in a controlled environment consisting of chairs, tables, laptops, staircases, etc. Figure 5 presents a top view of two obstacle courses where a user is required to move from a starting point to an end point along a path. The users had to walk slowly along a path since the feature extraction module was not fast enough. We found the BOVW computation module to be computationally expensive which required the user to be still for a while. The users were guided safely by the brief audio messages triggered by the SVM classifier immediately on classification of an object ahead. We observed that 'Paper Bins' and 'Staircases(Upper)' were detected and classified with 100% accuracy in test scenarios A and B. In a few cases, mis-classifications in case of the other objects left the user stranded. The matrices in Table 1 and 2 show that the classifier did get confused between 'door', 'laptop', and 'chair'. We observed a deterioration of performance in detection and classification with a reduction in lighting in the environment. We also noted that the users do not walk straight from the front towards an obstacle, they may approach from varying angles. Hence the angle at which Kinect RGB-D camera captures the images is crucial for a good field of view. We conclude from the confusion matrices in Table 1 and 2 that the feature extraction and classification algorithms worked satisfactorily well.

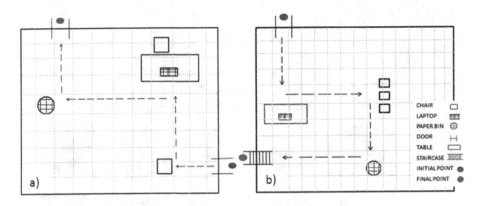

Fig. 5. Top view of two obstacle courses a) Test scenario A b) Test scenario B

As stated earlier, our environment was a controlled one (with fixed and limited indoor objects) and the navigation instructions for each route had to be coded into the classifier and was dependent on correct classification. Khan et. al. [13] have improvised on this and employed novel obstacle avoidance approaches to guide users. The multi-class SVM employed a *one-vs-one* strategy to classify multiple objects successfully. The audio feedback module was programmed to provide information about obstacles or staircases approximately 1-1.5 meters before the user reaches them, providing crucial time for a timely response. The different objects used in the test scenario are denoted in the legend in Fig. 5. The dashed arrow represents the direction along which the system prompted the blind-folded user to proceed.

Table 1. Confusion matrix of indoor objects seen in Test scenario A. Error cells in the bottom left indicate *false negatives* (2.5 %), while error cells in the top right indicate *false positives* (15.0%).

		Predicted Class				
		Chair	Laptop	Paper Bin	Door	Table
Actual Class	Chair	**7 (17.5%)**	2 (5.0%)	0 (0%)	0 (0%)	0 (0%)
	Laptop	1 (2.5%)	**6 (15.0%)**	0 (0%)	2 (5.0%)	1 (2.5%)
	Paper Bin	0 (0%)	0 (0%)	**8 (20.0%)**	0 (0%)	0 (0%)
	Door	0 (0%)	0 (0%)	0 (0%)	**6 (15.0%)**	1 (2.5%)
	Table	0 (0%)	0 (0%)	0 (0%)	0 (0%)	**6 (15.0%)**
	Total: 40, Errors: 7, *Accuracy*= 82.5%					

Table 2. Confusion matrix of objects seen in Test scenario B. Error cells in the bottom left indicate *false negatives* (5.4%), while error cells in the top right indicate *false positives* (5.35%).

		Predicted Class				
		Table	Laptop	Chair	Paper Bin	Staircase (Upper)
Actual Class	Table	**7 (12.5%)**	0 (0%)	0 (0%)	0 (0%)	0 (0%)
	Laptop	1 (1.8%)	**6 (10.71%)**	3 (5.35%)	0 (0%)	0 (0%)
	Chair	0 (0%)	2 (3.6%)	**21 (37.5%)**	0 (0%)	0 (0%)
	Paper Bin	0 (0%)	0 (0%)	0 (0%)	**8 (14.3%)**	0 (0%)
	Staircase (Upper)	0 (0%)	0 (0%)	0 (0%)	0 (0%)	**8 (14.3%)**
	Total: 56, Errors: 6, *Accuracy*= 89.31%					

5   Conclusion and Future Works

The traditional *white cane*, which is the standard navigation tool for the blind will be hard to replace because - it is cheaper, lighter, foldable and requires no power-source when compared to its modern competitors *i.e.* navigation assistive systems. *IntelliNavi* is a navigation assistive system that is meant to help the user gain improved understanding of the environment and thus complement the white cane and not necessarily replace it. We present this system which uses the Kinect sensor to sense the environment and learn indoor objects and deliver crucial information of the surrounding environment to the wearer through an external aid (earphone). An accurate labeling of the objects is done most of the time by the multi-class SVM trained with SURF descriptors and visual words. The experimental results show that our system can detect objects or obstacles with more than 82% accuracy consistently (which is comparable to [15] and [11]) and navigate a user successfully through predefined indoor scenarios in the presence of many real-world issues.

In spite of the promising results in experiments, several limitations of the system and considerations will need to be addressed. We observed that our application is not yet fully optimized. The BOVW module in the framework is found to be computational expensive, thus slowing up detection. This can be sped up with optimization. The multi-class SVM classifies multiple objects but detects only one object at a time. We plan to address these limitations and iteratively improve this application in future. The use of Computer Vision to recognize complex objects in cluttered scenes, detect people, localize natural or artificial landmarks, and thus assist in blind way-finding are promising research directions. Few researchers have worked on providing navigation to blind users in an outdoor environment [16]. The main challenge is to formulate a novel algorithm on navigation for indoor and outdoor environments.

The concept of this project was to test a technological concept that could, in the future, evolve into something more advanced. Obstacle detection and avoidance and blind navigation using Computer Vision and Machine Learning techniques is a very fertile field of research. Despite the prospect of increased independence enabled by assistive systems, we believe much advancement still needs to be made in terms of user-friendliness, portability, and practicality before such systems gain acceptance by the Visually Impaired community.

References

1. Caltech101 dataset, http://www.vision.caltech.edu/Image_Datasets/Caltech101/
2. Kinect for Windows, http://www.microsoft.com/en-us/kinectforwindows/
3. Visually Impairment and Blindness. WHO Fact sheet–282 (2014), http://www.who.int/mediacentre/factsheets/fs282/en/
4. Bay, H., Ess, A., Tuytelaars, T., Van Gool, L.: Speeded-Up Robust Features (SURF). Computer Vision and Image Understanding (CVIU) 110, 346–359 (2008)
5. Bernabei, D., Ganovelli, F., Benedetto, M.D., Dellepiane, M., Scopigno, R.: A Low-Cost Time-Critical Obstacle Avoidance System for the Visually Impaired. In: International Conference on Indoor Positioning and Indoor Navigation, Portugal, pp. 21–23 (September 2011)
6. Brock, M.: Supporting Blind Navigation using Depth Sensing and Sonification. In: ACM Conference on Pervasive and Ubiquitous Computing - UbiComp 2013, pp. 255–258. ACM, New York (2013)
7. Burges, C.J.C.: A Tutorial on Support Vector Machines for Pattern Recognition. Data Mining and Knowledge Discovery 2(2), 121–167 (1998)
8. Csurka, G., Dance, C.R., Fan, L., Willamowski, J., Bray, C., Maupertuis, D.: Visual Categorization with Bags of Keypoints. In: Workshop on Statistical Leanring in Computer Vision, pp. 1–22 (2004)
9. Eichhorn, J., Chapelle, O.: Object categorization with SVM: Kernels for Local Features. Tech. Rep. 137, Max Planck Institute for Biological Cybernetics, Tubingen, Germany (2004)
10. Filipe, V., Fernandes, F., Fernandes, H., Sousa, A., Paredes, H., Barroso, J.A.: Blind Navigation Support System based on Microsoft Kinect. Procedia Computer Science 14, 94–101 (2012)

11. Greenwood, C., Nirjon, S., Stankovic, J., Yoon, H.J., Ra, H.-K., Son, S., Park, T.: KinSpace: Passive Obstacle Detection via Kinect. In: Krishnamachari, B., Murphy, A.L., Trigoni, N. (eds.) EWSN 2014. LNCS, vol. 8354, pp. 182–197. Springer, Heidelberg (2014)
12. Hicks, S.L., Wilson, I., Muhammed, L., Worsfold, J., Downes, S.M., Kennard, C.: A depth-based head-mounted visual display to aid navigation in partially sighted individuals. PloS One 8(7), e67695 (2013)
13. Khan, A., Moideen, F., Lopez, J., Khoo, W.L., Zhu, Z.: KinDectect: Kinect Detecting Objects. In: Miesenberger, K., Karshmer, A., Penaz, P., Zagler, W. (eds.) ICCHP 2012, Part II. LNCS, vol. 7383, pp. 588–595. Springer, Heidelberg (2012)
14. Mann, S., Huang, J., Janzen, R.: Blind Navigation with a Wearable Range Camera and Vibrotactile Helmet. In: Proceedings of 19th ACM International Conference on Multimedia, pp. 1325–1328. ACM, New York (2011)
15. Sales, D., Correa, D., Osório, F.S., Wolf, D.F.: 3D Vision-Based Autonomous Navigation System Using ANN and Kinect Sensor. In: Jayne, C., Yue, S., Iliadis, L. (eds.) EANN 2012. CCIS, vol. 311, pp. 305–314. Springer, Heidelberg (2012)
16. Wang, S., Pan, H., Zhang, C., Tian, Y.: RGB-D Image-Based Detection of Stairs, Pedestrian Crosswalks and Traffic Signs. Journal of Visual Communication and Image Representation 25(2), 263–272 (2014)
17. Yang, J., Hauptmann, A.G.: Evaluating Bag-of-Visual-Words Representations in Scene Classification. In: International Workshop on Multimedia Information Retrieval (MIR 2007), pp. 197–206. ACM, New York (2007)
18. Zöllner, M., Huber, S., Jetter, H.-C., Reiterer, H.: NAVI – A Proof-of-Concept of a Mobile Navigational Aid for Visually Impaired Based on the Microsoft Kinect. In: Campos, P., Graham, N., Jorge, J., Nunes, N., Palanque, P., Winckler, M. (eds.) INTERACT 2011, Part IV. LNCS, vol. 6949, pp. 584–587. Springer, Heidelberg (2011)

A Trust Metric
for Online Virtual Teams and Work Groups

A.B. Sagar and Hrushikesha Mohanty

SCIS, Hyderabad Central University, Hyderabad,
India

Abstract. Online virtual teams publish their particulars on the Internet for collaborators and other stakeholders to make use of their services. It's usual that the stakeholders need to judge the level of trust they can put on a service provider before signing up an association. The process of building a dependent association with an Internet community suffers from usual drawbacks such as lack of proximity, anonymity, misrepresentation, proxy/masquerading, etc. Research of online cyber community and social networking, addresses some of these problems. These teams might be required to register with government agencies to add trust about their existence and functioning but stakeholders are usually in search of methods to verify trust on their own before availing services. This chapter introduces the concept of trust of online virtual teams and quantifies it. The subsequent sections define trust and identify its attributes and properties. They provide methods to quantify each attribute and finally, propose a metric for trust.

Keywords: online virtual teams, workgroups, trust, trust metric.

1 Introduction

Every business depends heavily on trust. Trust serves as a mechanism to reduce collaboration complexity [1]. Trust allows agents to resolve some of the uncertainty in their interactions, and form expectations about the behaviors of others. Good interactions between agents builds trust, which in turn allows future interactions to proceed as if certain unknown (or unknowable) quantities were known. While a great number of definitions of trust have been proposed in various disciplines, the present study defines trust intuitively as the belief that others will not harm or betray, that they will behave as expected, and that they will consider others' interests as if they were their own, when relied on them for some issue. Because of the facilitating role trust plays in multi-agent systems, much has been written on the subject in recent years. Many general-purpose trust models exist, and it is not the intention of the present work to replicate this body of existing work. Rather, the study focuses on dependability by making trust as a component of dependability. In this work, trust is between pairs of agents engaging in collaboration relationships, where tasks are transferred from one agent (the trustor) to another (the trustee) depending on the level of trust between them.

M.N. Murty et al. (Eds.): MIWAI 2014, LNAI 8875, pp. 184–195, 2014.

There are various trust models applied for solving different issues in literature. Some trust models are based on sound theories, e.g. PKI [2], some on reputations and recommendations [3] [4], and some on Probability or Bayesian network [5] [6]. Many trust models have been constructed for special computing environment such as ad hoc networks, peer-to-peer systems, and E-commerce [3] [4] [5] [6] [7] [8] [9] [10]. One of the most widely used definitions of trust in both human-organizational and multi-agent systems domains is given by Gambetta as, "Trust (or, symmetrically, distrust) is a particular level of the subjective probability with which an agent assesses that another agent or group of agents will perform a particular action, both before he can monitor such action (or independently of his capacity ever to be able to monitor it) and in a context in which it affects his own action " [11]. It is useful to mention this definition here, as it is generally well accepted in both human and multi-agent systems, and allows us to avoid negotiating the ongoing sociological debate on the precise meaning of trust.

Online virtual teams publish their particulars on the Internet for collaborators and other stakeholders to make use of their services. It's usual that the stakeholders need to judge the level of trust they can put on a service provider before signing up an association. The process of building a dependent association with an Internet community suffers from usual drawbacks such as lack of proximity, anonymity, misrepresentation, proxy/masquerading, etc. Research of online cyber community and social networking, addresses some of these problems. These teams might be required to register with government agencies to add trust about their existence and functioning but stakeholders are usually in search of methods to verify trust on their own before availing services. This chapter introduces the concept of trust of online virtual teams and quantifies it. The subsequent sections define trust and identify its attributes and properties. They provide methods to quantify each attribute and finally, propose a metric for trust.

1.1 Trust

Trust is primarily a cognitive issue and is perceived to vary from person to person and context. Research on trust has become necessary for activities on the Internet. In particular, the topic has become interesting for study because the trustee and trusted are not aware of each other and sometimes are located far apart. In such environments, computing trust has become a challenging problem. Considering these different problems, there have been several trust computing models. Before proposing a suitable trust model, let us touch upon the features that bear weight when making a model for computing the trust between teams.

Trust, being essentially a matter of cognition and abstraction, still having practical uses that need to be made tangible. The need is more in Internet based associations where there is an inherent difficulty among users because of anonymity and in-between remoteness. To be specific to the problem in hand, the proposed model for computing the trust of a team is based on three aspects — direct, indirect and recommender trusts.

The trustworthiness of a team is usually sensed from its activities and performance. A team that has been consistently working with perfection as per the specification is usually deemed the most trustworthy. Of course, another source for trust information about a team could be other users who had availed some service(s) earlier and also have something to say about the service providing team. And the third means to gather trust information on a team is from recommendations. A recommender could be thought of as an agency or expert committee that monitors team performance. They judge a team by observing its skills, professionalism, or scrutiny by visit. Before computing trust, let us study some of the properties of trust with respect to teams.

Properties of Trust. Some important properties of trust that are to be true in case of online teams [12] [13].

- Context dependence: A trust relation concerns a precise action or a particular entity and cannot be generalized to other actions or entities. In other words, trust is specific to a service being provided by a trustee. For example, team A may trust team B on a service such as *'providing raw material'*, but not on *'processing it'*. Thus, trust is specific to services that team offer in the context of a particular business.
- Measurability: Trust is also a measurable dimension of one's behavior. There can be several degrees of trust. For example, A may trust B more than C for a specific service. This property emphasizes the need for a suitable metric to represent trust. The measurement can be quantitative (e.g., a probability assignment) or relative (e.g., by means of a partial order).
- Dynamic: Trust is dynamic since it may evolve over a period of time. The fact that A trusted B in the past does not guarantee that A will continue to trust B in the future. B's behavior, performance and other relevant information may lead A to re-evaluate the trust on B from time to time. During a collaboration, the more A is satisfied with B's performance for service X, the more A trusts B. On the other hand, A's trust in B may decrease if B proves to be less trustworthy than anticipated by A.
- Sensitivity: Trust is very sensitive to malicious behavior. Trust declines very quickly with the detection of malicious behavior in a team. Once lost, a team needs many good interactions to regain a good level of trust.
- Asymmetry: A trust relation is asymmetric. That is, if A trusts B, it does not imply that B also trusts A equally. A may have high trust for B, but B may have much less trust or even distrust for A.
- Transitivity: Though controversial, in online teams, trust relation can be transitive. That is — if A trusts B, and B trusts C, then A can also trust C — to a certain level. But this is subject to constraints, and this property can hold true to a larger extent in teams. This property is essential for computation of indirect trust.

Context dependence property is satisfied by calculating trust with respect to a service. Measurability property is satisfied by attributing a quantitative value to the trust. Dynamic property is satisfied by calculating trust at random intervals. Sensitivity property is satisfied by defining penalties for malicious behaviors.

Asymmetry property is satisfied by computing trust with respect to each team. Transitivity property is satisfied by using recommendations from other teams and third parties.

Trust is to be calculated in three parts to satisfy all the above properties. That is, when team B calculates trust of team A, the first part is computed using the direct experiences with team A, the second part is obtained from associated teams which had direct experiences with team A, the third part is from the recommenders (expert committee). So, in essence, trust is of three forms — direct trust, indirect trust, and recommender trust.

Direct Trust Direct trust is based on self experiences and interactions with a target team. It is computed for behavioral features such as quality, speed of completion, cost effectiveness, and durability. Let us suppose that $team_i$ computes the direct trust value of $team_j$. If $team_j$ has interacted with $team_i$ before, then $team_i$ will have an opinion about the behavior of $team_j$. So, quantifying these behavioral features will help $team_i$ compute the direct trust of $team_j$. Let us represent direct trust value of $team_i$ on $team_j$ as $DT_{ij}^a(t)$, where 'a' represents the service and t represents the time at which trust is being calculated, since trust is time dependent and dynamic (a property of trust). As trust is also context dependent, trust is calculated for each service of $team_j$.

Trust is dynamic and therefore the trust value is to be updated each time a team encounters a situation that affects the trust value. When $team_j$ interacts with $team_i$ at time t, then $team_i$ updates its trust value for $team_j$ as follows:

$$DT_{ij}^a(t) = \begin{cases} DT_{ij}^a(t' + \delta t) + \Delta_{DT}(t) & \text{if } team_j \text{ is not new } team_i \text{ and } t' < t, \\ \Delta_{DT}(t) & \text{if } j \text{ is new to } i \end{cases}$$

Here, 't' represents time, and $t = t' + \delta t$. Here, δt represents the *'time elapsed'* since last update of trust value, and t' represents the time of last update. $\Delta_{DT}(t)$ represents the direct trust value that resulted due to direct experiences with $team_j$ at time t. On occasion, $\Delta_{DT}(t)$ could also be negative. Thus the trust measurement preserves the property of dynamic nature of trust [14].

Let us compute values for the components that comprise the direct trust. $DT_{ij}^{a.quality}$: This is the rating of team i to team j about the quality of service provided by j for the service/product/task a. Team i estimates $DT_{ij}^{a.quality}$ by verifying/observing/experiencing the quality provided by j at time t. Usually teams i and j make an agreement regarding the standard of quality that is to be maintained by team j. The range of values for quality component is $\{0,1\}$. If team j maintains quality standards and provides the service/product at an agreed quality, then $DT_{ij}^{a.quality} = 1$. Otherwise, if there was a compromise in the quality, then $DT_{ij}^{a.quality}$ value lies between 0 and 1. If the supplied quality is too inferior to the agreed quality, then $DT_{ij}^{a.quality} = 0$.

$DT_{ij}^{a.complSpeed}$: This is the value given by team i to team j about the speed of task execution by team j. This value is based on i's direct observation of j's performance. Usually teams i and j have an agreement regarding the speed of completion and deadlines to be followed. The range of values for complSpeed component is { 0,1 }. If team j adheres to the deadlines perfectly, then $DT_{ij}^{a.complSpeed} = 1$, otherwise, if the speed of completion is very slow, then $DT_{ij}^{a.complSpeed} = 0$. In other cases, it will be any intermediary value between 0 and 1.

$DT_{ij}^{a.costEffect}$: This is the value given by team i to team j about cost effectiveness of j with respect to service/task a at time t. This value is based on i's research on cost of 'a' quoted by j compared to other providers/executors. The range of values for costEffect component is { 0,1 }. If j has quoted less than the others, then the cost effectiveness of j is good, which is represented as $DT_{ij}^{a.costEffect} = 1$. If team j is found to be not cost effective, then $DT_{ij}^{a.costEffect} = 0$.

$DT_{ij}^{a.durability}$: This is the value given by team i to team j about durability of service/ product 'a' of j. This value is based on i's direct observation of j's product/service. Durability is the measure of life of a service/product. Durability does not always depend on quality. Sometimes one may use high quality components in making an item, but the durability of such an item may be much less. Mechanical systems characteristically fail from wear out mechanisms, while electronic systems more closely exhibit a constant failure rate over their useful life. Durability testing is the duration of time a product, part, material, or system can meet its performance requirements e.g. lifetime span. Depending on the type of material or product, team i simulates environmental factors or conditions to determine the operating limits or life span of the products, parts, and materials. Besides others, some of the tests are aging, operational limits, component degradation, failure points, material testing, tensile testing, burst testing, environmental testing, load testing, etc. The range of values for durability component is { 0,1 }. If the durability of an item a is high, then $DT_{ij}^{a.durability} = 1$, else, $DT_{ij}^{a.durability} = 0$.

Now that the values of all the attributes were quantified, $team_i$ can now compute direct trust value for $team_j$. Since the range of values for each attribute is {0, 1}, the range of values for direct trust will be {0, 4}.

$$DT_{ij}^a(t) = DT_{ij}^{a.quality}(t) + DT_{ij}^{a.complSpeed}(t) \\ + DT_{ij}^{a.costEffect}(t) + DT_{ij}^{a.durability}(t)$$

Indirect Trust. The second part of trust is indirect trust. This value is computed by an team for a target team with the help of the peers. A team can ask the peers to offer an indirect trust value for the target team. It is good if the peers are common friends to both the evaluating team and the target team. If such common peers are not available, then any teams that have direct experience(s) with the target team may be considered. Indirect trust value can be computed using feedback, associations, or social intimacy degree. Let us detail the strategies to calculate indirect trust from the behavioral features of teams.

Feedback. This is one of the methods for obtaining the indirect trust value of a target team. The peers of a target will be asked to give feedback for the target team. Indirect trust obtained from feedback can be represented as $IT_{ij}^{a.Feedback}(t)$ where IT represents Indirect Trust, 'a' represents the service/product for which Indirect Trust is being calculated, Feedback is the method employed, ij indicates that team i is computing Indirect Trust for team j. Feedback is usually taken for the following five aspects — cost, quality, transparency, timeliness and response time. As per [15], feedback for a team can be calculated as follows. Each aspect is measured on a 3-point scale calibrated as high, average, and low, and numerically each aspect may assume a value from set (1,2,3). For example, when cost is low, this aspect scores a value of 3; but for quality, low indicates a value of 1.

Let the components of feedback i.e. cost, quality, transparency, timeliness and response time, be represented by x_i. Then, feedback on team j to be obtained by team i for the service a can be computed from $\sum_{i=1}^{n} a.Feedback.x_i$. It is the sum of the scores of the components provided by the peers about the target team. Let us say that there are n components and let S_{mx} represent the 'maximum value of the scale' used to calibrate each satisfaction level. Currently, we considered only five aspects, each calibrated by a scale with S_{mx} value 3. Now we can compute the indirect trust value resulting from feedback using the below equation.

$$IT_{ij}^{a.Feedback} = \sum_{i=1}^{n} \frac{a.Feedback.x_i}{n * S_{mx}}$$

Associations. Feedback is not an '*always preferred*' approach. In society, trust on a person can be ascertained to a certain degree by judging the quality of his/her associations. Usually, a person is trusted if his/her associations are trusted [15]. Similarly, a team can infer an indirect trust value for a target team by looking at its associations for service/resource. So, it is possible for a team to choose alternate approaches to feedback due to circumstances, such as, peers not responding with feedback, the team does not have time to wait for the peers to respond, the team could not rely on the peers' responses, etc. One advantage of associations over feedback is the aspect of time. Inferring from associations has less delay compared to feedback. To compute feedback, a team needs to wait until the peers reply. But in the case of associations, there is no waiting time; a team itself can infer an indirect trust value by checking how good the associates of the target team are.

Let n be the number of associated teams, and CR_r^a represent the CollabRating value of a team r associated with team j for a service 'a'. Then, the association value that can be obtained by team i for team j can be represented as:

$$IT_{ij}^{a.Assocn} = \frac{\sum_{r=1}^{n}(CR_r^a)}{n}$$

1.2   Recommender Trust

The third part required for computing the total trust value for a team is *Recommender Trust*. In practice, recommendation has been a successful approach in building trust [15]. The team coordinator has a set of expert recommenders for each type of service in the online team system. These recommenders are experts in specific service/product domains they represent. When a team seeks recommendation regarding a target team, the recommenders check their data or use their custom methods to return a recommendation value for a target team. If $RT^a_{ij}(t) \in [0,1]$ indicates the recommender trust assigned to team j and revealed to team i for the service/product a, then $RT^a_{ij}(t) = 1$ implies that the recommenders fully recommend team j to team i regarding service a, and $RT^a_{ij}(t) = 0$ implies that the recommenders do not recommend team j to team i regarding service a. And, intermediate values show the degree of recommendation.

Trust Scale. On detailing the methods to compute the said three forms of trust, we can compute a trust value (T^a_{ij}) for a team as follows:

$$T^a_{ij} = DT^a_{ij} + IT^a_{ij} + RT^a_{ij}$$

Using the above computation, it is possible to arrive at a final trust value for a team with respect to a particular service/product a. The range of values for trust scale is $\{0,6\}$. This is because, the maximum value for IT^a_{ij} and RT^a_{ij} is 1, and for DT^a_{ij} is 4, and the minimum value for all three is 0. Now, $T^a_{ij} = 0$ implies that team j cannot be trusted for the service a, and $T^a_{ij} = 6$ means that the team j can be fully trusted for the service a. If it is any intermediate value, then level of trust for the specified service/product can be assumed accordingly.

1.3   Direct Trust Computation

To calculate direct trust for a service/product of a target team, we first need to compute values for quality, work speed, cost effectiveness, and durability.

Quality
Quality is assessed by a team for a target team's service/product. Quality evaluation can be made using a quality table. The attributes in the quality table are — Service/Product, team_id, Material Quality [MQ], Design Correctness [DC], Timeliness [TB], and Compliance to Standards [CoS]. Upon observation of the previously availed services from the target team, the evaluating team assigns values to each of these attributes of a service. Each attribute is given a 1 or 0 or any intermediate value between 0 and 1 such that 1 implies good/true and 0 for bad/false. The final value for quality is obtained by dividing the sum total of all the attribute values by 4.

$$\therefore Final\ quality\ value = \frac{MQ + DC + TB + CoS}{4}$$

Table 1. Quality assessment for Direct Trust

Service / Product (a)	Material quality [MQ]	Design correctness [DC]	Timeliness [T]	Compliance to standards [CoS]	Quality assessed
a	1	1	1	1	1
b	0	0	0	0	0

If a supplied service/product has good material quality, design correctness, timeliness, and compliance to standards, then MQ=1, DC =1, T=1 and CoS =1, and the quality value for that service/product is 1. But if a supplied service/product is poor in material quality, design correctness, timeliness, and compliance to standards, then MQ=0, DC =0, T=0 and CoS =0, making the quality value for that service/product to be 0. For intermediate values of these attributes, quality will also get an intermediate value between 0 and 1.

Work Speed

Based on the previous experiences, it is possible to compute work speed of a target team for a task using the following attributes — No. of days allotted for task completion, No. of days actually taken for completion, ratio of difference in days ($\pm$), Speed of Completion.

Table 2. Speed of Completion of Task

No. of days allotted for task completion (N)	No. of days actually taken for completion (M)	($\pm$) Delay ratio $\frac{(N-M)}{N}$	Speed of Completion
10	8	0.2	1
10	12	- 0.2	0.8
10	10	0	1

If $\frac{(N-M)}{N} = 0$, it means that the team has completed the task in perfect time and speed of completion = 1. If $\frac{(N-M)}{N} ¿ 0$, it means that the team has completed the task ahead of time and speed of completion = $1 + \frac{(N-M)}{N}$. But this is rounded to 1 as the maximum value for speed of completion is 1. If $\frac{(N-M)}{N} ¡ 0$, it means that the team has delayed the task execution and speed of completion = $1 - \frac{(N-M)}{N}$. The minimum value it can have is 0, since the minimum value of speed of completion is 0. For intermediate values of these attributes, work speed will also get an intermediate value between 0 and 1.

Cost Effectiveness

A team measures the cost effectiveness of a target team by comparing the cost claimed/levied by target team against the cost quotations of others. The attributes required for checking cost effectiveness are: Lowest cost quoted by other teams, cost quoted by target team, cost effectiveness, etc.

For a service, if the costs levied by the target team are more than what is quoted by the others, then the target team is not cost effective. Hence cost effectiveness in such a case is 0. However, if the target team quotes less than

Table 3. Cost effectiveness of service/product

Service/ Product	Lowest cost quoted by other teams (N)	Target team	Cost levied by target team (M)	Difference ratio $\frac{(N-M)}{N}$	Cost effectiveness
a	100	team j	80	0.2	1
b	100	team j	120	- 0.2	0
c	100	team j	100	0	0

the cost compared to other teams for the same service, then the target team is cost effective and the value of cost effectiveness is 1. For intermediate values of these attributes, cost effectiveness will also get some intermediate value between 0 and 1.

Durability

A team computes the durability of a target team's service/product by testing/observing the supplied service/product. To compute durability it uses a table with the following attributes — Service/Product, Type of Durability Test, Result, and Durability. Each service/product is tested with the relevant procedures.

Table 4. Durability of a product

Service/ Product	Durability Tests	Result	Durability
a	Pressure Testing, Thermal Testing, Load Testing	success	1
b	Aging, Burst Testing, Operational Limits	failure	0

Having computed values for Quality, Work Speed, Cost Effectiveness, and Durability of a service/product, Direct Trust for that service/product of the target team is computed by summation of all the four attributes.

1.4   Indirect Trust Computation

Indirect trust is computed with the help of the peers of the target team.

Feedback: Feedback is computed from the feedback values obtained from the peers of the target team. The obtained feedback values are fed into a feedback table. The attributes in the feedback table are — Servicename, teamId, Cost, Quality, Transparency, Timeliness, ResponseTime, Trust, and Timestamp. Each peer of the target team is sent a feedback form in which it has to respond to questions relating to quality, cost effectiveness, transparency, timeliness, and responsiveness of the target team. Each tuple in the feedback table represents the feedback by a team which has some direct trust value to the target team.

Feedback value of a target team can be computed as follows:

$$IT_{ij}^{a.Feedback}(t) = \sum_{i=1}^{n} \frac{a.FB.x_i}{n * S_{mx}}$$

$$= \sum_{i=1}^{n} \frac{a.FB.x_i}{15}$$

$FB.x_i$ indicates the attributes of the feedback table. S_{mx} represents the maximum value of the scale used for measuring each attribute, and n is the number of attributes. Since, $n = 5$ and $S_{mx} = 3$, we have $n * S_{mx} = 15$.

Associations

Computation of an association also needs the peers of the target team. For computation of association, a team uses the association table. The attributes in the table are — Target-team-id, ServiceName, Stakeholder1-CollabRating, Stakeholder2-CollabRating,.., AvgCollabRating. If a is the service, n is number of associated teams, then, association trust value of a target team is given by,

$$IT_{ij}^{a.Assocn}(t) = \frac{\sum_{i=1}^{n} a.CollabRating_i}{n}$$

Computation of trust due to association is not time consuming because it avoids waiting time for receiving feedback from all the peers; it only considers the Collaborativeness ratings of the peers and makes an average of it. Hence it is a time saving process.

1.5   Recommender Trust Computation

To compute recommender trust, a team uses the Recommenders Table which is managed by the coordinator or recommenders. It consists of information regarding all the expert recommenders nominated by the coordinator, and their areas of expertise. The attributes of the recommenders table are Recommenders, Service/Product, RecommendedTeam, RecommendationValue. The Recommendation value is given on the basis of the confidence of the recommenders and their discretion about the service/product of the target team. The highest value

Table 5. Feedback on the product

Service/ Product	Target teamId	Cost	Quality	Transp arency	Timeli ness	Response Time	Feedback giving team	Time stamp	Feedback Value
a	team098	3	3	3	3	3	team567	12-2-2012:04:50	1
a	team098	0	0	0	0	0	team433	22-2-2012:08:44	0

Table 6. Associations Trust

Service/ product	target teamId	Trusts of stakeholders	AssociatedTrust
a	team007	1,1,1	1
b	team007	0,0,0	0

for recommendation is 1, i.e. $RT_{ij}^a = 1$ and the lowest value is $RT_{ij}^a = 0$. Intermediate values carry the corresponding level of confidence in recommendation.

Table 7. Recommender Trust

Recommenders List	Service / Product	Recommended team	Location	Recommendation Value
rec_100, rec_34, rec_78	carpentry	team_007	Guntur	1
rec_120, rec_233, rec_484	garments	team_908	Hyderabad	0.5

2   Conclusion

This chapter introduced the concept of trust in teams and discussed its framework in the online team system. It also proposed a strategy for computation of trust of a team. The concept of trust arises when a team wants to collaborate with another team or when the coordinator wants to assign a task to a team. In both cases, verification is done for teams to determine if they are trustable or not. The properties and types of trust are discussed with relevance to teams. Trust, which is of three types — direct trust, indirect trust and recommender trust, is explained with a simple example. Also, the components of each type of trust are explained. Direct trust is computed from attributes such as quality, speed of completion, cost effectiveness and durability. Indirect trust is computed using one of the behavioural attributes such as feedback, associations, or social intimacy degree. Recommender trust is computed using recommendations of an expert committee. Competency and integrity are defined along with the processes of quantifying them have also discussed. The computation details for each of the attributes of direct trust, indirect trust, recommender trust were also given. The trust metric and its implications were also discussed. This chapter considered most of the attributes that have a significant influence on trust. It is possible to find even more attributes, but we feel that it is not required. The present attributes are good enough to determine whether a team is good enough for collaboration or not.

References

1. Lewis, J.D., Weigert, A.: Trust as a social reality. Social Forces 63(4), 967–985 (1985)
2. Perlman, R.: An overview of pki trust models. IEEE Network 13(6), 38–43 (1999)
3. Xiong, L., Liu, L.: A reputation-based trust model for peer-to-peer e-commerce communities. In: IEEE International Conference on E-Commerce, CEC 2003, pp. 275–284. IEEE (2003)
4. Michiardi, P., Molva, R.: Core: A collaborative reputation mechanism to enforce node cooperation in mobile ad hoc networks. In: Jerman-Blažič, B., Klobučar, T. (eds.) Advanced Communications and Multimedia Security. IFIP, vol. 100, pp. 107–121. Springer, Boston (2002)

5. Nguyen, C.T., Camp, O., Loiseau, S.: A bayesian network based trust model for improving collaboration in mobile ad hoc networks. In: 2007 IEEE International Conference on Research, Innovation and Vision for the Future, pp. 144–151. IEEE (2007)

6. Wang, Y., Vassileva, J.: Bayesian network trust model in peer-to-peer networks. In: Moro, G., Sartori, C., Singh, M.P. (eds.) AP2PC 2003. LNCS (LNAI), vol. 2872, pp. 23–34. Springer, Heidelberg (2004)

7. Theodorakopoulos, G., Baras, J.S.: Trust evaluation in ad-hoc networks. In: Proceedings of the 3rd ACM Workshop on Wireless Security, pp. 1–10. ACM (2004)

8. BenNaim, J., Bonnefon, J.F., Herzig, A., Leblois, S., Lorini, E.: Computer mediated trust in self interested expert recommendations. AI Soc. 25(4), 413–422 (2010)

9. Bonatti, P., Oliveira, E., Sabater-Mir, J., Sierra, C., Toni, F.: On the integration of trust with negotiation, argumentation and semantics. Knowledge Eng. Review 29(1), 31–50 (2014)

10. Resnick, P., Kuwabara, K., Zeckhauser, R., Friedman, E.: Reputation systems. Commun. ACM 43(12), 411–448 (2000)

11. Gambetta, D.: Can we trust trust. In: Trust: Making and Breaking Cooperative Relations, vol. 2000, pp. 213–237 (2000)

12. Abassi, R., El Fatmi, S.G.: Towards a generic trust management model. In: 2012 19th International Conference on Telecommunications (ICT), pp. 1–6. IEEE (2012)

13. Wen, L., Lingdi, P., Kuijun, L., Xiaoping, C.: Trust model of users' behavior in trustworthy internet. In: WASE International Conference on Information Engineering, ICIE 2009, vol. 1, pp. 403–406. IEEE (2009)

14. Bao, F., Chen, R.: Trust management for the internet of things and its application to service composition. In: 2012 IEEE International Symposium on World of Wireless, Mobile and Multimedia Networks (WoWMoM), pp. 1–6. IEEE (2012)

15. Mohanty, H., Prasad, K., Shyamasundar, R.: Trust assessment in web services: An extension to juddi. In: IEEE International Conference on e-Business Engineering, ICEBE 2007, pp. 759–762. IEEE (2007)

Web Service Composition Using Service Maps

Supriya Vaddi and Hrushikesha Mohanty

School of CIS, University of Hyderabad, India
{supriya_vaddi,hmcs_hcu}@yahoo.com

Abstract. Detailing a service structure may help to extend scope of a service search as well as composition. For the purpose we propose Service Map(SMap) for specifying a service. Also have defined a process for processing a query on SMap repository and shown that the process can also return a qualitative composed service, to meet query requirements. A framework as an extension to juddi registry, is proposed for implementation of SMap registry as well as processing of queries on the registry.

1 Introduction

A web Service is a self-describing, self-contained modular business application that can be published, located, and invoked over a network, such as Internet. Each service provides certain set of functionalities and these are exposed to the external world by interfaces for service search. The existing service search approaches are based on these interface descriptions e.g I/O match [6][3] and protocol match [5]. When a single service does not satisfy search request, a set of services are combined to meet the need. And, thus the process of obtaining a required service from a set of services is called web service composition.

Service composition helps in rapid development of applications, reuses the existing services, reduces cost of development and gives consumer a means to get a variety of composite services. The strategies most followed for service composition are: (i) search by function name (that corresponds to service functionality) (ii) following I/O match among services (iii) matching of service protocols, and these strategies work on interfaces that are exposed to external world. In our earlier work [11] we have argued a meta-description on service, i.e the way it's structured as well as promoted is useful for users to select a service of choice. For such a description we have proposed SMap a means to specify a service. The current work is an extension to the concept showing the usability as well as advantages of using SMap repository for service composition.

Here we propose a scheme that for a given query (specifying user request) finds services and composes them to a service that satisfies user queries. And a technique for query processing leading to service composition is detailed with a running example. Currently, it's being integrated with juddi[1] service engine. In order to defend registry based service composition, we would like to note that as searching for a service on web would be time consuming and may not result

M.N. Murty et al. (Eds.): MIWAI 2014, LNAI 8875, pp. 196–207, 2014.
© Springer International Publishing Switzerland 2014

in exact service that a user is expecting, hence registry based service search is tried out here.

This paper is organised as follows Section 2 details on existing service composition approaches, Section 3 gives an overview of SMap for service specification, syntax for querying SMaps and extensions supporting composition are discussed. In Section 4 approach for composing and ranking SMaps is discussed and algorithm is presented. An extension to juddi architecture supporting SMap specification & composition is shown in Section 5. We conclude with a remark in Section 6 .

2   Related Work

Generation of a composite service from a given set of specified service is an active area of research. Composition techniques and their success largely depends on the approaches specification follows. Here on reviewing some of the well known specifications and corresponding composition techniques the reasoning towards the proposed specification and the composition techniques will be presented.

The services are composed on matching the inputs and outputs [10] [9] [6] [7]based on function description [10] and sequence of message exchange [4] [8].

Different approaches graph [6] [9],Petrinet [10], Description logic [8],FSM [4] are used to generate composite services. To scale up the service composition RDBMS based approach is used in [7]. Below we would discuss each of these approaches in brief.

Composition of stateless services using dependency graph is presented in [6] Dependencies between input and output are represented as generalization and specialization, composition and decomposition in graph. Process algebra constructs sequence, choice, parallel, synchronization were introduced to represent the workflow of composition. A query with specific input and output would be processed by aligning a set of services in different combinations of introduced workflow.

Automatic service composition using logical inferences of horn clauses and petrinets is presented in [10]. The services are stored as propositional horn clauses capturing relationship between input and output (I/O) of the registered services.The query is modeled as the fact and goal in the Horn clauses. Then the service composition problem is transformed into the logical inference problem of Horn clauses. An algorithm is presented to determine the entailment of the query and to select the clauses necessary for inference. Petri nets are chosen to model the selected Horn clause set and clients query (fact and goal statement).

To enhance service composition semantic annotations are given to I/O parameters in [9]. Proposed composition algorithm takes services specified in OWL-S as input and builds a semantic network on applying similarity measures equivalent, subsumed by and independent on I/O. For a given query semantic graph is traversed to find composite service with output similar to goal.

The paper [8] proposes a process based ontological framework for service composition supporting orchestration and choreography of services using description logic. A service process requirement is described in description logic with combinators sequence, parallel, non-deterministic and iteration on set of service functions. Process expressions describing service choreography uses different notations for send and receive actions thus enabling composition of choreographed services. Framing simple queries is easier but as the complexity of query increases it is difficult for a user to specify a query in description logic.

In paper [4] the services are described as execution trees comprising of actions performed when a service is executed. These trees are converted to state machines and set of services meeting the user query are selected on matching published FSM's with that of queried FSM.

To scale up service compositions [7] stores service data in relational database. In this paper service composition solutions are precomputed on building a composition graph (using joins) and storing all the paths in a table. For a given query a part of the path is searched for the requested input-output and all such matching paths are retrieved from graph. Additionally ontology information is stored for each parameter based on the degree of match and to aid query processor in retrieving similar services.

Majority of works for composition are based on matching I/O, process, function-name and message exchange sequence of services with that of query. selected services are chained in different forms sequence, parallel, choice to form a composite service. But a service query may be for granular service functionalities and oriented towards promotions. To meet queries of this nature SMap specification was introduced to describe services to granular level and to be compatible with existing specification standards (WSDL) communicating point details are given to select appropriate service composition.

Research works[7][6][9] discussed above have precomputed composition graphs where as our approach of service composition is memory based, we would initially use data base tables to find requested service items and their detail, then the communicating point details are obtained from WSDL repository and these CPTs are matched for compatibility to get composed service.

3   SMap for Specification and Composition

A user in need of a service would have some basic idea of functionalities expected. Query syntax for querying over SMaps allows consumer to specify query on processing certain functionalities as query items retrieves a set of SMaps containing these items. When an SMap does not contain all the query items, in that case SMaps that collectively contain all query items are to be composed to a single service for meeting user needs.

Let there be a service ABCRestaurant with a service item socialfunction that includes conferencehall and XYZFlorist service with a service item bookFlowers providing various ArrangementPattern including WeddingFlowers.

A user in need of conference hall and wedding flowers would query for these two query items in service repository. As ABCRestaurant contains conference hall but not wedding flowers the query is partially satisfied similar is the case with XYZFlorist service. Hence these services are to be composed to fully satisfy the user query. If a user has specific constraints like the wedding flowers are to be booked only after conference hall is booked, then a service has to communicate its state with the next service.

So, a query decides whether intended result needs a composition of services as a service or not. The issues involved for composing with SMaps are (a)communication details are missing in existing specification (b)the order in which services are to be composed is not known. This paper addresses these issues and an approach to achieve composition of services specified by SMaps is discussed.

3.1 SMap Specification and Its Extension

In our previous works [12] [11] we have proposed SMap for service specification and querying. SMap is a web service specification to detail design of a service with different service items and the way these items are packaged to cater to different needs of users. Each service item is connected to the other using associations viz.includes(inc), a-kind-of (akf), leads-To (lst) and has-with (hw). The first association 'inc' is used to specify the internal structure of service that details on containment of an item in another. When one or more items are providing similar kind of functionalities then these items are alternative to each other and are connected by an edge labelled 'akf'. Promotions on items are connected with edge labelled 'lst'. Qualitative details on service items with different attributes are specified using 'hw' construct. When two or more items are provided in combinations *and* operator is used to specify constraints on item togetherness, *or* operator specifies mutual exclusiveness among service items.

The SMap syntax to specify a service is given below in BNF notation

$WSMap := < ServiceName > < Association > <ItemNode>*$
$<ItemNode> := <ItemDetail> [<Association> | <Operation><ItemNode>]*$
$<ItemDetail> := <ItemName>[has<FeatureName>[with <FeatureValue>]\ ]$
$< Association > := <Functional> | < Causal >$
$<Functional > := "inc" | "akf"$
$< Causal > := "lst" <PromoCond>$
$<Operation > := "and" | "or"$

Thus, a service can be packaged in a modular fashion so that even each module can be invoked independently. This enables a consumer to avail a service by part even. Further work, we show this adds to service flexibility enabling a service resilient by making a module replaceable by another similar module available else where.

For the purpose SMap specification is extended by declaring a node in SMap as communicating point CP (a CP node is double oval shaped refer Fig. 1). The nodes reachable from CP make a functional module. A CP is labelled by I/O parameters telling the inputs it requires at the time of invocation, and return it makes at exit.(refer. Fig.1). This (I/O parameters) make interface to a SMap module and this needs to be exposed to external world for interested service consumers. In [11] we have specified schemas that store SMaps in a RDBMS.

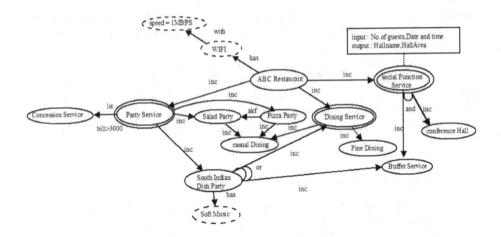

Fig. 1. Hotel Service specified in SMap annotated with CP's

Now, along with these schemas information CP's are to be stored. For this we define a new schema CPT. Below, we present all the schemas for storing SMaps.

  ServiceItemDetail :: (<u>SId</u>, <u>ItemName</u>, FeatureName, FeatureValue)

  Includes :: (<u>SId</u>, <u>ItemName</u>, IncludedItems, CombiType)

  LeadsTo :: (<u>SId</u>, <u>ItemName</u>, PromoItem, PromoCond)

  AKindOf :: (<u>SId</u>, <u>ItemName</u>, AlternativeItem)

  SMap :: (<u>SId</u>, ServiceName, PS)

  CPT :: (<u>SId</u>,<u>ItemName</u>,Input,Output)

Multiple communicating points can be present in an SMap each present at different levels from root service item. CP present at each level takes different inputs and returns different outputs based on the included sub functionalities.

The next section illustrates the use of this extended schema of SMap repository for the use of service composition.

3.2   Querying SMaps - A Composition Perspective

Earlier in [11] we have discussed on mechanism to specify and find a service that suffices a query. Here, we'll take up the case of a query that needs services from more than one SMap. The core idea of processing such a query rests on two issues: (i) First according to requested query items identifying the corresponding SMaps and then (ii) Verify the composability of chosen SMaps. One possible solution to the composability is the order service items are required by a service user. And during query processing it's to be ascertained whether a specific order (in a request) is maintainable or not. If it is, then services are composable and for the request, a composed service is achievable. Before giving detail on query processing we deal on query specification as follows

GetS with Sitem (<servItem> |(<servItem> **and|before|after** <servItem>)
[and(<servItem> **has** <servFeat>)|(<servItem> **has** <servFeat> **with** <featValue>)
[**withItemDetail** (<servItem> **inc** <servItem>)]
[**withPromo** (<promo> **on** <servItem> **with** <servCond>)]
[**withAltTo**<servItem>]

A user, in a query specifies the order in which user wishes to avail the services named in *servItems*. Thus, essentially a workflow schedule is defined on a query. Further, the syntax has a provision to detail on a service item by associating its features, feature value and business promotions. It can also specify a construct of service item by specifying what it includes of. Further, alternative to a service item can also be specified. With reference to our previous work [12], the specification proposed here specifies service workflow schedule with the help of *and*, *before* and *after* operators, that having their usual meanings.

4 Service Composition : A Process

There are two methods of composing services - static and dynamic. In static service composition, the services that are to be composed are selected and address bounded in advance where as in dynamic service composition the services are bounded during service execution. In this paper we address the problem of static service composition i.e to identify a set of best services that are compatible to each other. The composition is performed in three stages. In stage one SMaps containing matching query terms are identified, in second stage the composability is checked and in the third stage the composed services are ranked based on their features. Each of these stages are discussed below in detail.

4.1 Identifying Service Maps

Revisiting the example from section 3 and re-framing the query following the query syntax - "*GetS with Sitem conferencehall before wedding flowers*"

If the published services in SMap repository are ABCRestaurant service with service modules diningservice, socialfunction service and party service at the first level. Similarly XYZFlorist Service is published with SMap as shown in left half of Fig.2, sub functionalities viz. arrangementPattern, bookFlowers, orderTracking and delivery are accessed though two CP's *bookFlowers* and *ordertracking*.

To process the above query initially the SMap repository is searched for SMaps containing conferenceHall and WeddingFlowers this would retrieve ABCRestaurant and XYZFlorist services ref. Fig.2, and these services are to be composed.

For a repository containing more than one SMaps with the query items then the result would be sets of SMaps for each query term. Let m_1 be set of Smaps containing query term q_1(conferencehall) then the set would contain $S_{11}, S_{12}, S_{13}....S_{1m_1}$ SMaps. Similarly m_2 be set of Smaps containing query term q_2(weddingflowers) then the set would contain $S_{21}, S_{22}, S_{23}....S_{2m_2}$ SMaps. Now an SMap from each of these sets is selected to check for compatibility. And finally the selected SMaps S_1 and S_2 are to be composed.

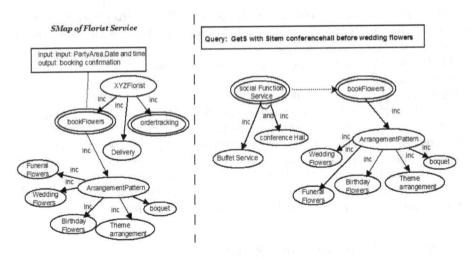

Fig. 2. SMap of XYZ Florist service with CP's and

A query item can be present at any location in SMap. Hence a path running through the requested query item is to be obtained known as *service thread*. The next sub section discusses the selection of service threads from SMaps for composition.

4.2 Composability of SMaps

The retrieved services are composed on identifying part of SMap (*member service*) that contains the query term and compatibility among the identified parts are checked.

Member Service is a part of a SMap that contains query items and spans from their youngest ancestor node with a communicating point, to all reachable leaf nodes of the SMap.

This part of SMap is a candidate service that is deliverable and can be integrated to another service for making of a composed service. A communicating point on a SMap is labelled with input and output parameters pertaining to the member service it hosts. This provision in SMap brings in granularity in a service description making flexible for a user to choose and pay for the exact service of need, and enables service provider to offer a variety of service provisioning, in a service catering to the types of needs of users.

Service composition by Input Output match is a well studied and is itself a topic of research. Here, our focus being on flexible service specification we dont deal on I/O match based service composition in detail. Rather we concentrate on process of finding number of services from SMaps for service composition.

For a given query with q_i a set of query items say 'n', the search on SMap registry may return several member services 'ms'. It may happen for a query item q_i, we may get a set of service maps (Fig.3) S_{ij} with 'm' selections. So, we need to choose the best on ranking defined on it. When two query items q_i and q_j are of different member services say S_x and S_y with communication points CP_x and CP_y then for service composition we look for the match.

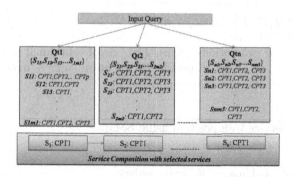

Fig. 3. Pictorial representation of options for services selection

Two CP's CP_x and CP_y are said to be compatible with each other when output of a CP_x exactly matches the input of CP_y. An algorithm addressing service composition is presented in section 4.3.

Going back to query shown in Section 4.1, Conference hall is included in socialfunction service, and hence this part of SMap is selected as member service. And for the second query item 'weddingflowers' it is included in 'arrangement-Pattern' which in turn is exposed though CP 'bookFlowers' a member service as shown in Fig.2. Next step is to check the compatibility between the services i.e to identify the communicating points and to match them according to the order specified in query. Here the communicating point detail as obtained from SMaps:

SocialFunction

   input : No.of guests, Date and time   *output* : Hallname, HallArea

XYZFlorist

   input: PartyArea, Date and time   *output*: BookingConfirmation

The output of social function service (HallArea) matches with the input of florist service (PartyArea) and the second argument data and time is known from user, hence they are compatible to be composed.

   The query can also be descriptive as *getS with Sitem SocialFunctionService before FloristService with ItemDetail SocialFunctionService inc buffet service* then the query processor has to work on the internals of SMap to match the sub requirements and then proceed with composition.

4.3   Service Composition and Ranking

When many pairs of SMaps are found compatible then there exists several composition options but to select the best composition ranking of ms is important.

Ranking of ms is done following preference score of member service (PS_{ms}) computed based on member service features. Composition of top scored ms's have more promotional, qualitative features and less conditions hence they are preferred for composition.

$$PS_{ms} = (P + F_Q - C) * (n_{ms} / n_{ms}^{max})$$

Where P- no. of promotions in ms, F_Q - no.of additional quality features, C-no.of conditions, n_{ms} - no.of included items in ms, n_{ms}^{max} - maximum number of included items for an ms belonging to set m_i.

Algorithm 1. Service Composition using SMaps

1: **Input:** ConsumerQuery Q, SMapRepository;
2: **Output:** composite service
3: Array qts,FullMatch,matched[][],composedsrv;
4: **Begin**
5:    qts←getQueryTerms(Q) {***Step :1***}
6:    FullMatch←getServiceswithAllQTs(qts);
7:    **if** FullMatch is not NULL **then**
8:        Display FullMatch Services
9:    **else**
10:       **for** each $qt_i \in$ qts **do**
11:           $m_i =$ getServiceCountWithQT(qt_i)
12:       **end for**{***Step:2***}
13:       **for** Each qt_i in qts **do**
14:           **for** Each S_x in m_i and S_y in m_{i+1} **do**
15:               **if** CP_out(ms[S_x]) = CP_in(ms[S_y]) **then**
16:                   **if** CP(ms[S_x]) not in matched[i] **then**
17:                       matched[i] = CP(ms[S_x])
18:                   **end if**
19:                   **if** CP(ms[S_y]) not in matched[i+1] **then**
20:                       matched[i+1]= CP(ms[S_y])
21:                   **end if**
22:               **end if**
23:           **end for**
24:       **end for**{***Step:3***}
25:       **for** k=1;k< |qts|;k++ **do**
26:           **if** matched[k]≠ null and matched[k+1]≠null **then**
27:               op = getoperator(k,k+1)
28:               composedsrv = matched[k] op matched[k+1]
29:           **end if**
30:       **end for**
31:    **end if**
32: **End**

Algorithm to compose SMaps is shown in Algorithm.1 it takes consumer query and SMap repository as input and gives composed service (set of services to be composed) as output. Ranking of services is not presented in the algorithm due to space constraint.

The steps in composing SMaps are *Step1*: To check composition need, *Step2*: To retrieve set of compatible services *Step3*: generate service compositions.

The algorithm internally uses a set of functions viz. getQueryTerms takes query as input and gives set of query terms as output and getServiceswithAllQTs takes the query terms as input and retrieves an SMap that contains all the query terms. getServiceCountWithQT takes query term as input and retrieves number of services in repository containing this query term. CP takes member service as input and retrieves the communication point detail as output.

Step1: Initially the service items are searched in repository if all the query items are found in an SMap then *FullMatch* is assigned with matched service ids and are listed (LineNo: 5-7) if there doesn't exist any such service then Sids containing each query term qt are retrieved Line No: 10-11.

Step2: for each service(S_x,S_y), in the retrieved list the member services are identified ($ms[S_x]$, $ms[S_y]$) and the communicating points associated with each ms ($ms[S_x]$) is matched with that of member service of next SMap ($ms[S_y]$) line: 13-26 if they match and the matched SMap is not present in the matched list then it is added else the next SMap is searched for compatibility match.

Step3: From lines 27-32 the operator is added to the selected services for composition, the operator can be 'sequence' or 'parallel'. The obtained communicating points can be composed using BPEL [2] to generate and deploy an executable composite service.

5   Framework for SMap Composition Extending Juddi

This section gives an overall framework for service specification, querying and composition using SMaps. The framework shown in Fig.4 is an extension to existing juddi registry architecture. The framework allows service provider to publish services using *SMap management* module, which gives provider an UI to create, store and edit SMaps.

The published SMaps are validated as per the specification syntax before storing. SMap Preprocessor parses the diagrammatic information and extracts the SMap detail to store in SMap Data base. The communication detail specified in SMap is stored as t-model in juddi. When a user queries for a service, modules under *SMap Query Processor*, process the query. Initially query is checked for syntax correctness in *QueryPreProcesor*. Later the query items are searched in SMap database if all the items are present in an SMap then all the matching services are displayed to user in decreasing order of their preference using *ServiceRanker*, Where as for other case the services are composed by *Service composer*. It has *MSSelector* module to identify the member services from each SMap meeting the user query. Later these services are checked for the compatibility of communicating points using *CompatibilityChecker*. The list of thus obtained services are aligned, ranked internally based on Preference Score of member services discussed in Section 4.3 and best services are selected and composed by *CompositeServiceGenerator*. The proposed work addresses composition

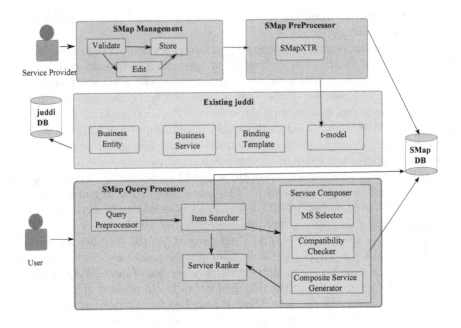

Fig. 4. Service Composition

of stateless services, In our future work we would address composition of stateful services and dynamic service composition using SMaps.

6   Conclusion

Registry based Service Oriented Architecture is at times useful for making a large number of services available at one place to a service user. We have proposed a technique, that on applying to a repository of SMaps, produces composite service to match a query of user interest: SMaps, the specification of web services are stored in relational database. For queries, a SQL-like syntax is proposed. Processing a query on SMap service repository leads to a composite service. In this paper we have detailed on query processing technique and outlined its working with an example. Further we have proposed a framework for possible implementation of the technique. Before, [11] we have extended juddi enabling it to store SMaps in a relational database. With this experience we are now implementing our proposed query processor for SMap repository at juddi.

Acknowledgements. This work is sponsored by UGC under Basic Research Fellowship in Sciences (BSR) for meritorious students vide UGC letter No. F.7-337/2011(BSR) dated 14.3.2014.

References

1. Java uddi (juddi) implementation. Apache Project
2. Ws-bpel specification. Web services business process execution language version 2.0 oasis standard (2007)
3. Aversano, L., Canfora, G., Ciampi, A.: An algorithm for web service discovery through their composition. In: Proceedings of the IEEE International Conference on Web Services, pp. 332–339 (July 2004)
4. Berardi, D., Calvanese, D., De Giacomo, G., Lenzerini, M., Mecella, M.: Automatic composition of E-services that export their behavior. In: Orlowska, M.E., Weerawarana, S., Papazoglou, M.P., Yang, J. (eds.) ICSOC 2003. LNCS, vol. 2910, pp. 43–58. Springer, Heidelberg (2003)
5. Corrales, J.C., Grigori, D., Bouzeghoub, M.: BPEL processes matchmaking for service discovery. In: Meersman, R., Tari, Z. (eds.) OTM 2006. LNCS, vol. 4275, pp. 237–254. Springer, Heidelberg (2006)
6. Hashemian, S.V., Mavaddat, F.: A graph-based framework for composition of stateless web services. In: Proceedings of the European Conference on Web Services, ECOWS 2006, pp. 75–86. IEEE Computer Society, Washington, DC (2006)
7. Lee, D., Kwon, J., Lee, S., Park, S., Hong, B.: Scalable and efficient web services composition based on a relational database. Journal of Systems and Software 84(12), 2139–2155 (2011)
8. Pahl, C., Zhu, Y.: A semantical framework for the orchestration and choreography of web services. Electronic Notes in Theoretical Computer Science 151(2), 3–18 (2006), Proceedings of the International Workshop on Web Languages and Formal Methods (WLFM 2005)
9. Talantikite, H.N., Aissani, D., Boudjlida, N.: Semantic annotations for web services discovery and composition. Computer Standards & Interfaces 31(6), 1108–1117 (2009)
10. Tang, X., Jiang, C., Zhou, M.: Automatic web service composition based on horn clauses and petri nets. Expert Systems with Applications 38(10), 13024–13031 (2011)
11. Vaddi, S., Mohanty, H.: Querying a service from repository of SMaps. In: Hota, C., Srimani, P.K. (eds.) ICDCIT 2013. LNCS, vol. 7753, pp. 535–546. Springer, Heidelberg (2013)
12. Vaddi, S., Mohanty, H., Shyamasundar, R.K.: Service maps in xml. In: Proceedings of the CUBE International Information Technology Conference, CUBE 2012, pp. 635–640. ACM, New York (2012)

Integrated Representation of Spatial Topological and Size Relations for the Semantic Web

Sotiris Batsakis, Grigoris Antoniou, and Ilias Tachmazidis

Department of Informatics
University of Huddersfield
Queensgate, Huddersfield, West Yorkshire, HD1 3DH, UK
{S.Batsakis,G.Antoniou,Ilias.Tachmazidis}@hud.ac.uk

Abstract. Representing topological information for the Semantic Web often involves qualitative defined natural language terms such as "In". This can be the case when exact coordinates of spatial regions are not available or they are unreliable. Topological spatial relations are the most important aspect of spatial representation and reasoning. In addition to topology, often the exact size of regions is not known as well, but the relations between sizes (e.g., larger, equal) are known. In this work we propose and evaluate an integrated representation of topological relations (RCC-5 and RCC-8) and qualitative size relations using OWL object properties and axioms, combined with reasoning rules expressed using SWRL, embedded into the ontology. Different representations are implemented and evaluated.

Keywords: Spatial Representation and Reasoning, Semantic Web, Rules.

1 Introduction

Semantic Web technologies are used for automating tasks handled manually by users, tasks such as reasoning over data and answering complex questions. Understanding the meaning of data, typically in RDF[1] format, requires formal definitions of concepts and their properties, using the Semantic Web Ontology definition language OWL [5]. OWL allows for reasoning over the definitions and the assertions of concepts and specific individuals using reasoners such as HermiT [9] and Pellet [10]. Furthermore, reasoning rules can be embedded into the ontology using the SWRL [4] rule language.

Spatial information can be defined using quantitative (e.g., using coordinates) and qualitative terms (i.e., using natural language expressions such as "In"). Qualitative spatial terms have specific semantics which can be embedded into the ontology using reasoning rules. In [1], such a representation, embedding semantics by means of SWRL rules, was proposed for spatial and temporal information in OWL. Specifically, directional and topological information (based on the RCC-8 [3] set of topological relations) were represented (in [11,12] similar representations for topologic relations were proposed as well) , but size information was not represented as this work does.

---

[1] http://www.w3.org/standards/techs/rdf

M.N. Murty et al. (Eds.): MIWAI 2014, LNAI 8875, pp. 208–219, 2014.

Embedding reasoning rules into the ontology makes sharing of data easier since all SWRL compliant reasoners (such as Pellet and HermiT) can be used for spatial reasoning. To the best of our knowledge, this work is the first that represents both topological and size relations for the Semantic Web, using representations based on the direct integration of topological (both RCC-5 and RCC-8) and size relations. In addition to that, representations based on the decomposition of topological and size relations into two basic sets, reasoning over each set and an interaction model between the two sets using SWRL are proposed as well. It is also the first work that deals with combined spatial (both topologic and size) reasoning based on a rule based approach embedded into OWL ontologies. Furthermore, tractable sets for all representations and combinations of relations are identified and presented.

Current work is organized as follows: related work in the field of spatial knowledge representation is discussed in Section 2. The proposed combined representation (for RCC-5 and size relations) is presented in Section 3 and the corresponding reasoning mechanism in Section 4. The representation of RCC-8 and size relations is presented in Section 5 followed by evaluation in Section 6 and conclusions and issues for future work in Section 7.

2 Background and Related Work

Definition of ontologies for the Semantic Web is achieved using the Web Ontology Language OWL. The current W3C standard is the OWL 2 [5] language, offering increased expressiveness while retaining decidability of basic reasoning tasks. Reasoning tasks are applied both on the concept and property definitions into the ontology (TBox), and the assertions of individual objects and their relations (ABox). Reasoners include among others Pellet[2] and HermiT[3]. Reasoning rules can be embedded into the ontology using SWRL[4]. *Horn Clauses* (i.e., a disjunction of clauses with at most one positive literal), can be expressed using SWRL, (i.e., $\neg A \vee \neg B... \vee C$ can be written as $A \wedge B \wedge ... \Rightarrow C$).

Qualitative spatial reasoning (i.e., inferring implied relations and detecting inconsistencies in a set of asserted relations) typically corresponds to Constraint Satisfaction problems which are *NP-hard*, but tractable sets (i.e., solvable by polynomial time algorithms) are known to exist [2,8].

Embedding spatial reasoning into the ontology by means of SWRL rules applied on topologic relations forms the basis of the SOWL model proposed in [1,7], but the problem of combining topological and size relations was not addressed. Such information is required for representing expressions such as for example " regions R3 and R2 overlap with region R1, and R2 is larger than the other two regions". Current work addresses this issue by applying the bi-path consistency method [2,8]. Furthermore, by embedding reasoning rules into the ontology, the proposed representation offers greater flexibility since it can be used and modified freely using only standard Semantic Web tools such as

---

[2] http://clarkparsia.com/pellet/
[3] http://hermit-reasoner.com/
[4] http://www.w3.org/Submission/SWRL/

the Protégé editor and the HermiT reasoner. In addition to that, tractable sets identified in this work can be used for optimizing the performance of spatial CSP solvers [2].

3 Spatial Representation

Topological RCC-5 relations in this work are represented as object properties between OWL objects representing regions. For example, if $Region1$ is In $Region2$, the user asserts the binary relation $Region1$ PP (proper part) $Region2$, or equivalently $PP(Region1, Region2)$. This is also the approach used for size relations. This approach is similar to the approach used in [1]. The first representation proposed in this work implements reasoning rules applied on integrated topological and size relations. Following this approach for example, if a region overlaps and has equal size with another region, then this is represented by asserting one relation between them. The second approach is based on decomposing relations, for example if a region overlaps and has equal size with another region, then two relations (one topologic and one size) are asserted as object properties.

The proposed representation deals with size (qualitative size or QS) relations between regions in addition to their topologic (RCC) relations. The combined relation is abbreviated as RCC-QS. Between two regions three possible size relations can hold, these relations are "$<$", "$>$", "$=$" also referred to as *smaller, larger* and *equals* respectively.

Region Connection Calculus [3] is one of the main ways of representing topological relations. There are several variants of the calculus corresponding to different levels of detail of the represented relations, variants such as RCC-5 and RCC-8. In the following the representation and reasoning of RCC-5 relations is presented.

RCC-5 relations is a set of 5 topological relations namely DR (discrete), PO (partially overlapped), EQ (equals), PP (proper part) and PC (contains). Figure 1 illustrates these relations between two regions X and Y. Relations DR, PO and EQ are symmetric, and relation PP is the inverse of PC. All these 5 basic RCC-5 relations are pairwise disjoint. Also EQ, PP and PC are transitive. All the above can be represented using OWL object property axioms (i.e., symmetry, inverse, disjointness and transitivity).

The integrated RCC-QS representation consists of the following basic relations:

$$DR\_smaller(x, y) \equiv smaller(x, y) \land DR(x, y)$$
$$DR\_equals(x, y) \equiv equals(x, y) \land DR(x, y)$$
$$DR\_greater(x, y) \equiv greater(x, y) \land DR(x, y)$$
$$PO\_smaller(x, y) \equiv smaller(x, y) \land PO(x, y)$$
$$PO\_equals(x, y) \equiv equals(x, y) \land PO(x, y)$$
$$PO\_greater(x, y) \equiv greater(x, y) \land PO(x, y)$$
$$EQ\_equals(x, y) \equiv equals(x, y) \land EQ(x, y)$$
$$PP\_smaller(x, y) \equiv smaller(x, y) \land PP(x, y)$$
$$PC\_greater(x, y) \equiv greater(x, y) \land PC(x, y)$$

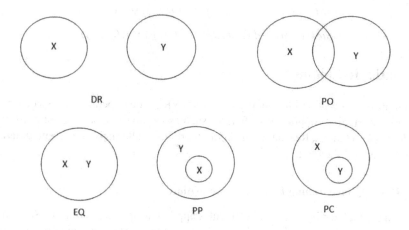

Fig. 1. Topological RCC-5 Relations

Notice that relation PP can be combined only with size relation "$<$", PC only with "$>$" and EQ with "$=$", while PO and DR can be combined with all size relations. In total the combined $RCC\text{-}QS$ relations set consists of 9 basic relations and several additional relations representing disjunctions of basic relations that are needed for supporting reasoning rules of Section 4.

3.1   Decomposition of RCC-QS Relations

Reducing the complexity of the previous representation can be achieved by translating $RCC\text{-}QS$ relations to two sets of relations, that need fewer rules to reason with, but also more property assertions. Between two regions, one topological relation of Figure 1 (or a disjunction of these basic relations of Section 4) and one size relation (*smaller, greater or equal*) can be asserted. Axioms between these relations, based on their semantics, are part of the representation (e.g., DC, PO, EQ are symmetric PO,PC are inverses of each other and *smaller* is inverse of *greater*). Additional SWRL rules are required for representing property restrictions propagated from one set of relations to the other:

$$PP(x,y) \rightarrow smaller(x,y)$$

$$EQ(x,y) \rightarrow equal(x,y)$$

$$PC(x,y) \rightarrow greater(x,y)$$

These rules enforce that; when a region is a proper part of another it must also be smaller than the other, if it is identical to the other then their sizes are equal, and if it contains another region then its size must be greater than that of the other region, respectively. Additional restrictions represented using relations representing disjunctions (see Section 4) are also part of the representation:

$$smaller(x,y) \rightarrow DR(x,y) \vee PO(x,y) \vee PP(x,y)$$

$$greater(x, y) \rightarrow DR(x, y) \lor PO(x, y) \lor PC(x, y)$$

$$equals(x, y) \rightarrow DR(x, y) \lor PO(x, y) \lor EQ(x, y)$$

4   Spatial Reasoning

Reasoning is realized by introducing a set of SWRL rules operating on spatial relations. Reasoners that support DL-safe rules such as HermiT can be used for reasoning over *RCC-QS* relations. Reasoning rules for all proposed alternative representations are presented in the following.

4.1   RCC5-QS Reasoning Using Decomposition

Reasoning is realized by a set of SWRL rules applied on spatial relations of Section 3. Defining compositions of relations is a basic part of the spatial reasoning mechanism. Table 1 represents the result of the composition of each pair of topological RCC-5 relations of Figure 1.

Table 1. Composition Table for RCC-5 Topological Relations

Relations	DR	PO	EQ	PP	PC
DR	All	DR,PO,PP	DR	DR,PO,PP	DR
PO	DR,PO,PC	All	PO	PO,PP	DR,PO,PC
EQ	DR	PO	EQ	PP	PC
PP	DR	DR,PO,PP	PP	PP	All
PC	DR,PO,PC	PO,PC	PC	PO,EQ,PP,PC	PC

The composition Table can be interpreted as follows: if relation R_1 holds between *Region*1 and *Region*2 and relation R_2 holds between *Region*2 and *Region*3, then the entry of the Table 1 corresponding to line R_1 and column R_2 denotes the possible relation(s) holding between *Region*1 and *Region*3. For example, if *Region*1 is *Proper Part* (*PP*) of *Region*2 and *Region*2 is *Proper Part* (*PP*) of *Region*3 then *Region*1 is *Proper Part* (*PP*) of *Region*3. Entries in the composition table are determined using the formal semantics of Region Connection Calculus as defined in [3].

Relations between regions representing sizes are *smaller, greater* and *equals*, denoted as "<",">","=" respectively. Table 2 illustrates the set of composition reasoning rules of existing size relation pairs.

Reasoning over spatial relations is known to be an *NP-hard* problem and identifying tractable cases (i.e., sets of relations that a polynomial algorithm is sound and complete) of this problem has been in the center of many research efforts over the last few years [2,1]. The notion of *k-consistency* is very important in this research. Given a set of n regions, with relations asserted between them imposing certain restrictions, *k-consistency* means that every subset of the n regions containing at most k regions does not contain an inconsistency. Note that, checking for all subsets of n entities for consistency is exponential on n. Reasoning over the 5 basic RCC-5 relations and 3 basic size relations

Table 2. Composition Table for qualitative size relations

Relations	<	=	>
<	<	<	<, =, >
=	<	=	>
>	<, =, >	>	>

can be done using path-consistency, which is a polynomial time method, since these basic relations belong to tractable sets [2,1].

A series of compositions of relations may yield relations which are inconsistent with existing ones (e.g., inferring using compositions for example, that X *into* Z will yield a contradiction if X *overlaps* Z has been also asserted into the ontology). Consistency checking is achieved by ensuring path consistency by applying formula:

$$\forall x, y, k \; R_s(x, y) \leftarrow R_i(x, y) \cap (R_j(x, k) \circ R_k(k, y))$$

representing intersection of compositions of relations with existing relations (symbol $\cap$ denotes intersection, symbol $\circ$ denotes composition and R_i, R_j, R_k, R_s denote topological relations). The formula is applied until a fixed point is reached (i.e., the application of the rules above does not yield new inferences) or until the empty set is reached, implying that the ontology is inconsistent. Implementing path consistency formula requires rules for both compositions and intersections of pairs of relations. Compositions of relations R_1, R_2 yielding a unique relation R_3 as a result are expressed in SWRL using rules of the form:

$$R_1(x, y) \wedge R_2(y, z) \rightarrow R_3(x, z)$$

The following is an example of such a composition rule:

$$PP(x, y) \wedge DR(y, z) \rightarrow DR(x, z)$$

Rules yielding a set of possible relations cannot be represented directly in SWRL, since disjunctions of atomic formulas are not permitted as a rule head. Instead, disjunctions of relations are represented using new relations, whose compositions must also be defined and asserted into the knowledge base. For example, the composition of relations PO and PP yields the disjunction of two possible relations (PP and PO) as a result:

$$PO(x, y) \wedge PP(y, z) \rightarrow (PO \vee PP)(x, z)$$

If the relation PO_PP represents the disjunction of relations PO and PP, then the composition of PO and PP can be represented using SWRL as follows:

$$PO(x, y) \wedge PP(y, z) \rightarrow PO\_PP(x, z)$$

A set of rules defining the result of intersecting relations holding between two regions must also be defined, in order to implement path consistency. These rules are of the form:

$$R_1(x,y) \wedge R_2(x,y) \to R_3(x,y)$$

where R_3 can be the empty relation. For example, the intersection of relations DR and PC yields the empty relation ($\perp$ or $null$), and an inconsistency is detected:

$$DR(x,y) \wedge PC(x,y) \to \perp$$

Intersection of relations PO and PO_PP (representing the disjunction of $Overlaps$ and *Proper Part*) yields relation PO as a result:

$$PO(x,y) \wedge PO\_PP(x,y) \to PO(x,y)$$

Thus, path consistency is implemented by defining compositions and intersections of relations using SWRL rules, and OWL axioms for inverse relations as presented in Section 3. Another important issue for implementing path consistency is the identification of the additional relations, such as the above mentioned PO_PP relation, that represent disjunctions. Specifically, the *minimal* set of relations required for defining compositions and intersections of all relations that can be yielded when applying path consistency on the basic relations of Figure 1 is identified. The identification of the additional relations is required for the construction of the corresponding SWRL rules.

Table 3. Closure method

Input:Set S of tractable relations
Table C of compositions
WHILE S size changes
BEGIN
Compute C:Set of compositions of relations in S
S=S $\cup$ C
Compute I:set of intersections of relations in S
S= S $\cup$ I
END
RETURN S

In this work, the *closure method* [2] of Table 3 is applied for computing the minimal relation sets containing the set of basic relations: starting with a set of relations, intersections and compositions of relations are applied iteratively until no new relations are yielded, forming a set closed under composition, intersection and inverse. Since compositions and intersections are constant-time operations (i.e., a bounded number of table lookup operations at the corresponding composition table is required) the running time of closure method is linear to the total number of relations of the identified set. Furthermore, tractability of the initial set guarantees tractability of the new set as well [2].

Applying the closure method over the set of basic RCC-5 relations yields a set containing 12 relations. These are the 5 basic relations of Figure 1 and the relations DR_PO representing the disjunction of DR and PO, DR_PO_PC representing the disjunction of DR, PO and PC, DR_PO_PP representing the disjunction of DR, PO and PP, $PO\_EQ\_PP\_PC$ representing the disjunction of PO, EQ, PP and PC, PO_PP representing the disjunction of PO and PP, PO_PC representing the disjunction of PO and PC, and All denoting the disjunction of all relations. Path consistency and closure method are also applied on size relations using the compositions defined on Table 2.

In addition to path consistency and closure method, interactions between the two sets must be taken into account. These interactions are represented using the rules of Section 3.1. These rules are also used when detecting minimal tractable sets using the closure method, and this modified version of path-consistency/closure using interaction rules is characterized as bi-path consistency[2]. Applying bi-path consistency over both sets of relations yields a tractable set consisting of 18 relations (14 RCC-5 and 4 size relations).

4.2   Reasoning over Integrated RCC-5 and Size Relations

Instead of using two relations between two regions, one for topology and the other for size, one relation representing the combined information can be used instead (e.g., *contains-and-greater*). These combined relations are defined in Section 3. Compositions are defined using the composition table for each set of relations and combining the results. For example:

$$PC\_greater(x,y) \circ PC\_greater(y,z) \equiv$$
$$(PC(x,y) \circ PC(y,z)) \wedge (greater(x,y) \circ greater(y,z)) \equiv PC(x,z) \wedge greater(x,z)$$
$$\equiv PC\_greater(x,z)$$

After defining the combined composition table, path consistency and the closure method are applied on the set of basic relations of Section 3. Since there is one set of relation integrating topological and size relations, additional rules representing interactions between sets as in Section 3.1 are not required. In total 27 basic and disjunctive relations are required for the combined $RCC5\text{-}QS$ representation.

5   Reasoning over RCC-8 Topologic and Size Relations

Another important set of topological relations is the RCC-8 set of relations, which is a refinement of RCC-5 relations set. Specifically, the DR relation is refined into two distinct relations; DC (Disconnected) representing the fact that two regions do not have common points, and EC (Externally connected) representing the fact that two regions have common boundary points, but not common internal points. Similarly the PP (Proper part) relation is refined into two different relations TPP and $NTPP$. TPP is representing the fact that a region is a proper part of another region and also has common points with the boundary of the enclosing region. $NTPP$ on the other hand, represents the fact that the enclosed region does not have common points with the boundary of the enclosing region. $NTPP_i$ and TPP_i are the inverses of $NTPP$ and TPP respectively. RCC-8 relations are illustrated in Figure 2.

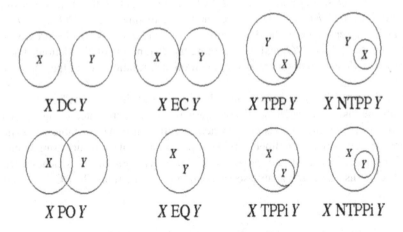

Fig. 2. Topological RCC-8 Relations

The composition table for RCC-8 relations has been defined in [3], and an implementation based on path consistency using SWRL and OWL axioms was presented in [1]. This is the first work dealing with combined RCC-8 and size information (*RCC8-QS*) for the semantic Web. Size relations are identical to those that were combined with RCC-5 relations, as presented in Sections 3.1 and 4.1.

Two different representations (combined with the corresponding reasoning mechanisms) are proposed. The first is similar to the representation of Section 3. Relations *DC,EC,PO* are compatible with all three size relations yielding 9 valid combined relations. Relations *TPP, NTPP* are compatible only with *smaller* (2 valid relations), *TPPi,NTPPi* with *greater* (2 valid relations) and *EQ* with *equals* (1 valid relation). In total 14 relations are needed for representing combinations of basic RCC-8 and size relations. Path consistency and the closure method are applied directly over these relations as in Section 4. After applying the closure method, the identified minimal tractable set contains 82 relations. Alternatively, two relations can be asserted, combined with rules imposing the above mentioned compatibility restrictions. Restrictions are implemented using SWRL rules as in Section 3.1 and reasoning is achieved using bi-path consistency of Section 4.1. Applying this method, a tractable set containing 56 relations (52 RCC-8 and 4 size relations) is identified.

6   Evaluation

Reasoning is achieved by employing DL-safe rules expressed in SWRL, that apply on named individuals in the ontology ABox, thus retaining decidability. Furthermore, using the integrated *RCC-QS* representation, reasoning using the polynomial time path-consistency algorithm is sound and complete, as in the case of the decomposed representation using bi-path consistency [2,8].

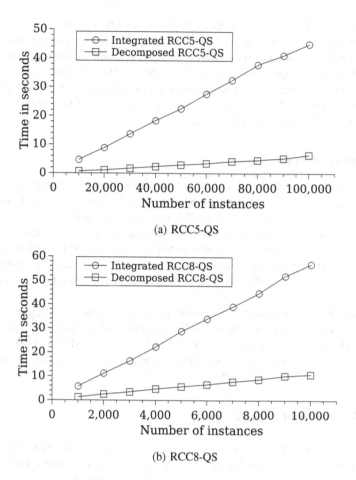

Fig. 3. Average reasoning time for (a) RCC5-QS (b) RCC8-QS relations

Specifically, any region can be related with any other region with two basic relations, one topological and one size (or one relation in case of the integrated representations). Since relations of each set are mutually exclusive, between n regions, at most $2n(n-1)$ relations can be asserted. Furthermore, path-consistency has $O(n^5)$ time worst case complexity [6](with n being the number of regions). In the most general case where disjunctive relations are supported in addition to the basic ones, any region can be related with every other region by at most k relations, where k is the size of the set of supported relations (containing for example, 27 relations in case of $RCC5\text{-}QS$ algebra using the integrated approach). Therefore, for n regions, using $O(k^2)$ rules, at most $O(kn^2)$ relations can be asserted into the knowledge base.

The $O(n^5)$ upper limit for path consistency running time referred to above is obtained as follows: At most $O(n^2)$ relations can be added in the knowledge base. At each such addition step, the reasoner selects 3 variables among n regions (composition rules) which corresponds to $O(n^3)$ possible different choices. Clearly, this upper bound is pessimistic, since the overall number of steps may be lower than $O(n^2)$ because an inconsistency detection may terminate the reasoning process early, or the asserted relations may yield a small number of inferences. Also, forward chaining rule execution engines employ several optimizations (e.g., the Rete algorithm employed at the SWRL implementation of Pellet), thus the selection of appropriate variables usually involves fewer than $O(n^3)$ trials. Nevertheless, a worst case selection of variables can be assumed in order to obtain an upper bound for complexity.

6.1 Experimental Evaluation

Measuring the efficiency of the proposed representation requires a spatial ontology of Section 3 using both approaches (integrated and decomposition) containing instances and relations (both *RCC5-QS* and *RCC8-QS*), thus a synthetic data-set of regions generated randomly was used for the experimental evaluation. Reasoning response times of the spatial reasoning rules are measured as the average over 10 runs. HermiT 1.3.8 running as a library of a Java application was the reasoner used in the experiments. All experiments where run on a PC, with Intel Core CPU at 2.4 GHz, 6 GB RAM, and Windows 7.

Measurements illustrate that the proposed approach can efficiently represent thousands of regions and reason over them in a few seconds without using any specialized software besides a standard OWL reasoner such as HermiT. In addition to that, decomposition based representations are much more efficient in terms of reasoning speed, in case of both *RCC5-QS* and *RCC8-QS*.

7   Conclusions and Future Work

In this work, a representation framework for handling topological (both RCC-5 and RCC-8) and size relations in ontologies is introduced. The proposed framework handles both, size and RCC relations using an inference procedure based on path consistency. Two representations are proposed, an integrated representation, and a representation based on decomposition of RCC-QS relations.

The proposed representations are fully compliant with existing Semantic Web standards and specifications, which increases their applicability. Being compatible with W3C specifications, the proposed framework can be used in conjunction with existing editors, reasoners and querying tools such as Protégé and HermiT, without requiring specialized software. Therefore, information can be easily distributed, shared and modified. Directions of future work include the development of applications based on the proposed mechanism. Furthermore, rule based reasoning is suitable for parallel implementations, which is a promising direction of future research.

References

1. Batsakis, S., Petrakis, E.G.M.: SOWL: A Framework for Handling Spatio-Temporal Information in OWL 2.0. In: Bassiliades, N., Governatori, G., Paschke, A. (eds.) RuleML 2011 - Europe. LNCS, vol. 6826, pp. 242–249. Springer, Heidelberg (2011)

2. Renz, J., Nebel, B.: Qualitative Spatial Reasoning using Constraint Calculi. In: Handbook of Spatial Logics, pp. 161–215. Springer, Netherlands (2007)
3. Cohn, A.G., Bennett, B., Gooday, J., Gotts, N.M.: Qualitative spatial representation and reasoning with the region connection calculus. Geoinformatica 1(3), 275–316 (1997)
4. Horrocks, I., Patel-Schneider, P.F., Boley, H., Tabet, S., Grosof, B., Dean, M.: SWRL: A Semantic Web Rule Language Combining OWL and RuleML. W3C Member submission (2004), http://www.w3.org/Submission/SWRL/
5. Motik, B., Patel-Schneider, P.F., Horrocks, I.: OWL 2 Web Ontology Language: Structural Specification and Functional-Style Syntax. W3C Recommendation (2009), http://www.w3.org/TR/owl2-syntax/
6. Batsakis, S.: Reasoning over 2D and 3D directional relations in OWL: A rule-based approach. In: Morgenstern, L., Stefaneas, P., Lévy, F., Wyner, A., Paschke, A. (eds.) RuleML 2013. LNCS, vol. 8035, pp. 37–51. Springer, Heidelberg (2013)
7. Christodoulou, G., Petrakis, E.G.M., Batsakis, S.: Qualitative Spatial Reasoning using Topological and Directional Information in OWL. In: Proc. of 24th International Conference on Tools with Artificial Intelligence (ICTAI 2012), November 7-9 (2012)
8. Gerevini, A., Renz, J.: Combining topological and size information for spatial reasoning. Artificial Intelligence 137(1), 1–42 (2002)
9. Shearer, R., Motik, B., Horrocks, I.: HermiT: A Highly-Efficient OWL Reasoner. In: OWLED, vol. 432 (2008)
10. Evren, S., Parsia, B., Cuenca Grau, B., Kalyanpur, A., Katz, Y.: Pellet: A practical owl-dl reasoner. Web Semantics: Science, Services and Agents on the World Wide Web 5(2), 51–53 (2007)
11. Marc-Zwecker, S., De Bertrand De Beuvron, F., Zanni-Merk, C., Le Ber, F.: Qualitative Spatial Reasoning in RCC8 with OWL and SWRL. In: KEOD 2013-International Conference on Knowledge Engineering and Ontology Development (2013)
12. Batsakis, S., Antoniou, G., Tachmazidis, I.: Representing and Reasoning over Topological Relations in OWL. In: Proceedings of the 4th International Conference on Web Intelligence, Mining and Semantics (WIMS 2014), p. 29. ACM (2014)

Using Bayesian Networks to Model and Analyze Software Product Line Feature Model

Musfiqur Rahman and Shamim Ripon

Department of Computer Science and Engineering,
East West University, Dhaka, Bangladesh
m.rahman.2014@ieee.org, dshr@ewubd.edu

Abstract. Proper management of requirements plays a significant role in the successful development of any software product family. Application of AI, Bayesian Network (BN) in particular, is gaining much interest in Software Engineering, mainly in predicting software defects and software reliability. Feature analysis and its associated decision making is a suitable target area where BN can make remarkable effect. In SPL, a feature tree portrays various types of features as well as captures the relationships among them. This paper applies BN in modeling and analyzing features in a feature tree. Various feature analysis rules are first modeled and then verified in BN. The verification confirms the definition of the rules and thus these rules can be used in various decision making stages in SPL.

Keywords: Software Product Line, Bayesian Networks, False Optional, Dead feature.

1 Introduction

Software Product Line (SPL) is a set of related softwares, also known as software family, where the member products of the family share some common features and each member is characterized by their varying features [4]. The main objectives behind SPL is reusability, time to market, increased quality [3,4]. The common and varying features of a SPL are arranged in model that helps the stakeholder to select their required product feature configuration. Common requirements among all family members are easy to handle as they simply can be integrated into the family architecture and are part of every family member. But problem arises from the variant requirements among family members as modeling variants adds an extra level of complexity to the domain analysis. Thus management of variants is considered to be one of the critical areas in SPL.

Constructing a feature model that correctly represents the intended domain is a critical task [2] and defects may be introduced while constructing the feature model. If defects are not identified and corrected at proper stage, it can diminish the expected benefit of SPL [10]. Among the various types of defects found in feature model, *dead* and *false optional* features are two most common types of defects that are of our interest in this paper. A dead feature is a feature

M.N. Murty et al. (Eds.): MIWAI 2014, LNAI 8875, pp. 220–231, 2014.

that is part of a feature tree but it never appears in any valid product [9,10]. Having a dead feature in FM indicates inaccurate representation of the domain requirements. A false optional feature on the other hand is a feature that is declared as an optional feature but it appears in all valid products. Both dead and false optional feature arise due to misuse of dependencies among features.

Due to increased complexity and uncertainly in software requirements, application of AI techniques are becoming evident for managing them [12]. Bayesian Network (BN) [7,8,13] has already been successfully applied to various areas in Software Engineering [6,11,14,15,19] as well as in Requirement Engineering [1]. Besides, a probabilistic feature model is presented in [5] by showing its use in modeling, mining and interactive configuration. Although various work have been carried out where BNs have been used to predict software defects, reliability, etc., very few addressed the requirement management issues in SPL, in particular feature model analysis. Our focus in this paper is to address this area where the inference mechanism and conditional probability in BNs can be used to analyze various defects in feature model.

The objective of this paper is to adopt AI technique to analyze the defects in SPL feature models. This paper focuses on two specific analysis operations: *false optional* and *dead features* of feature diagram. By extending our earlier definition [17], we define several analysis rules for false optional and dead features by using First Order Logic (FOL). We then represent these rules by using BNs. First, we define BNs graphs to represent the scenarios of the our analysis rules. After defining node probability tables of the variables of BNs graphs, the conditional probability of the variables related to analysis rules are calculated. The calculated results match our FOL analysis rules.

The rest of the paper is organized as follows. Section 2 gives a brief review of feature tree and BN and shows how to draw BN of a feature tree. Section 3 defines First Order Logic based analysis rules of false optional and dead features. Section 4 models the analysis rules in BN and then show the probabilistic calculation for the features. Finally, in Section 5 we conclude the paper by summarizing our contributions and outlining our future plans.

2   Feature Diagram and Bayesian Network

Feature modeling is a notation proposed in [9] as a part of domain analysis of system family. Feature modeling facilitates addressing the common and varying features of a product family in both formally and graphically. In feature model features are hierarchically organized as trees where root represents domain concepts and other nodes represents features. Features are classified as *Mandatory* and *Optional*. The relationship among the child sub-features of a parent feature are categorized into *Or* and *Alternative*. When there is a relationship between cross-tree (or cross hierarchy) features (or variation points) we denote it as a dependency.

BN is a directed graph where the nodes represent variables (events), and the arrows between the nodes represent relationship (dependencies). Each node in the graph has an associated set of Conditional Probability Distributions (CPD). For discrete variables, the CPD can be represented as a Node Probability Table (NPT) which list the probabilities that given node takes on each of its different values of each combination of values of its parents. When the nodes have two possible values, binary probabilities (0 or 1) are used in CPD. The construction of a BN starts with the identification of the relevant variables in the domain that has to be modeled. After that the dependencies between the variables has to be determined. Finally CPD is added for each variable.

A BN also corresponds to a directed acyclic graph (DAG). The directed edges represent the direction of dependencies between the nodes. A directed edge from node X_i to X_j indicates that X_i is an ancestor of X_j and the value of X_j depends on the value of X_i. Feature tree on the other hand is a rooted tree, where non-root nodes represent features, and features (parent) and sub-features (child) are connected via edges. Due to the strong resemblance between BNs and feature trees, BNs can be conveniently used to represent the information within feature tree. Added with this, BNs can also include probabilistic information to support decision making in uncertain situations. Figure 1 shows an example feature tree and its corresponding BN representation. Based on the dependency information of the the variable, a NPT can be constructed for each variable in the BN.

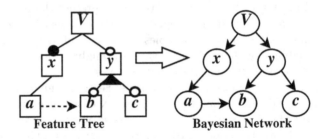

Fig. 1. BN representation of a feature tree

3   Feature Modeling Analysis Rules

Defects in feature model have adversary effects on the quality of the derived product model. Dead and false optional features are two most common types of defects found in any FM. A dead feature is a feature that is part of a feature tree but it never appears in any valid product [9,10]. Having a dead feature in FM indicates inaccurate representation of the domain requirements. A false optional feature on the other hand is a feature that is declared as an optional feature but it appears in all valid products. Both dead and false optional feature arise due to misuse of dependencies among features. Extending our earlier work [17] on logical representation of feature model and their analysis rules, we present in this section a set of rules for dead and false optional features. Our definitions are also influenced by those mentioned in [16].

We use the following FOL predicates to define the rules.

Table 1. Summary of Analysis Rules

	Scenario	Analysis Rules
False Optional		$\forall v, x, y \cdot variation\_point(v) \wedge mandatory\_variant(v, x)$ $\wedge optional\_variant(v, y) \wedge requires(x, y)$ $\wedge select(x) \Rightarrow select(y)$
		$\forall v, x, y, z \cdot variation\_point(v) \wedge mandatory\_variant(v, x)$ $\wedge optional\_variant(v, y) \wedge optional\_variant(v, z)$ $\wedge requires(x, y) \wedge requires(y, z)$ $\wedge select(x) \Rightarrow select(y) \wedge select(z)$
		$\forall v, x, y, z \cdot variation\_point(v) \wedge mandatory\_variant(v, x)$ $\wedge optional\_variant(v, y) \wedge variation\_point(y)$ $\wedge optional\_variant(y, z) \wedge requires(x, y)$ $\wedge select(x) \Rightarrow select(y) \wedge select(z)$
Dead Features		$\forall v, x, y \cdot variation\_point(v) \wedge mandatory\_variant(v, x)$ $\wedge optional\_variant(v, y) \wedge excludes(x, y)$ $\wedge select(x) \Rightarrow \neg select(y)$
		$\forall v, x, y, z \cdot variation\_point(v) \wedge mandatory\_variant(v, x)$ $\wedge optional\_variant(v, y) \wedge optional\_variant(v, z)$ $\wedge requires(x, y) \wedge excludes(y, z)$ $\wedge select(x) \Rightarrow select(y) \wedge \neg select(z)$
		$\forall v, x, y, z \cdot variation\_point(v) \wedge mandatory\_variant(v, x)$ $\wedge optional\_variant(v, y) \wedge variation\_point(y)$ $\wedge optional\_variant(y, z) \wedge excludess(x, y)$ $\wedge select(x) \Rightarrow \neg select(y) \wedge \neg select(z)$
		$\forall v, x, y, z \cdot variation\_point(v) \wedge mandatory\_variant(v, x)$ $\wedge optional\_variant(v, y) \wedge optional\_variant(v, w)$ $\wedge variation\_point(y) \wedge optional\_variant(y, z)$ $\wedge excludess(x, y) \wedge requires(w, z) \wedge select(x)$ $\Rightarrow \neg select(y) \wedge \neg select(z) \wedge \neg select(w)$

- *variation_point(v)*: This predicate indicates that feature v is a variation point, i.e., feature v has child feature(s).
- *mandatory_variant(v, x)*: This predicate indicates that feature x is a mandatory feature of feature v, i.e., x is a mandatory child of v.
- *optional_variant(v, x)*: Here, x is an optional feature of v.
- *requires(x, y)*: It indicates that feature x requires feature y.
- *exclude(x, y)*: This predicate indicates that feature x and y are mutually exclusive.
- *select(x)*: This predicates the selection of a variant x.

The three analysis rules are defined for false optional features and four analysis rules for dead features. Table 1 illustrates the rules written using FOL. Later these definitions will be used during the analysis of Bayesian Networks based definitions.

4   Analysis Rules in BN

After analyzing the scenarios for false optional and dead features in a feature tree, we define Bayesian network representation of the scenarios. The key factor in defining the BN presentation is to identify the dependencies among the features. Our BN is heavily influenced by our logical representation shown in [17]. Table 2 illustrates the BN of the analysis rules mentioned earlier.

Table 2. BN representation of false optional and dead feature analysis rules

False Optional, Rule 1: An optional feature becomes false optional when a mandatory feature requires that optional feature. In the Bayesian network (1st rule of false optional in Table 2), both x and y are two variants of the variation point v. Thus the probability of both of these features being selected are directly dependent on v. Feature x depends only on v, whereas y depends on both x and v. The NPT of the variable v, x and y is given in Fig. 2.

v			x			y				
T	F		V	T	F	V	x	T	F	
1	0		T	1	0	F	F	0	1	
			F	0	1	F	T	I	I	
						T	F	I	I	
						T	T	1	0	

Fig. 2. NPT for variables in false optional rule 1

In the NPT, I stands for inconsistent. From the NPT the probability of selecting the variable y can be calculated by using Bayesian rule,

$$P(y, x, v) = P(y \mid x, v)P(x \mid v)P(v)$$

$$P(y = T \mid x = T) = \frac{P(x = T, y = T)}{P(x = T)} = \frac{\sum_{v \in \{T,F\}} P(x = T, y = T, v)}{\sum_{v \in \{T,F\}} P(x = T, v)}$$

$$= \frac{P(y = T, x = T, v = T) + P(y = T, x = T, v = F)}{P(x = T, v = T) + P(x = T, v = F)}$$

$$P(y = T, x = T, v = T) = P(y = T \mid x = T, v = T)P(x = T \mid v = T)P(v = T)$$
$$= 1 \times 1 \times 1 \times 1 = 1$$

Similarly, the probabilities of other variables can also be calculated. Finally we get the following conclusion,

$$P(y = T \mid x = T) = \frac{1 + 0}{1 + 0} = 1$$

The result indicates that the optional variant y will be always selected whenever the variant x is selected. As x is a mandatory feature, it will always be selected and thus y will be always part of a valid product, even though it is declared as optional. Hence, y is a false optional feature.

False Optional, Rule 2: An optional feature becomes false optional when it is required by another false optional feature. In the 2nd rule for false optional in Table 2, x is a mandatory feature which requires an optional feature y. Thus y becomes a false optional feature. Moreover, y also requires an optional feature z which implies that z is also a false optional feature. As the three variants x, y, and z are the variants of v, in BN all of these variables are directly dependent of v. Besides, y has a dependency on v and x and z has on v and y. Even there is no direct dependency between x and z, variant z is selected whenever x is selected. The NPT of the variables v, x, y and z are given in Fig 3.

From the NPT the probability of selecting the variable z can be calculated by using Bayesian rule,

v	
T	F
1	0

x		
v	T	F
T	1	0
F	0	1

y			
v	x	T	F
F	F	0	1
F	T	I	I
T	F	I	I
T	T	1	0

z			
v	y	T	F
F	F	0	1
F	T	I	I
T	F	I	I
T	T	1	0

Fig. 3. NPT for false optional rule 2

$$P(z, y, v) = P(z \mid y, v) P(y \mid x, v) P(x \mid v) P(v)$$

$$P(z = T \mid y = T) = \frac{P(z = T, y = T)}{P(y = T)} = \frac{\sum_{x,v \in \{T,F\}} P(z = T, y = T, x, v)}{\sum_{x,v \in \{T,F\}} P(y = T, x, v)}$$

$$= \frac{\begin{array}{c} P(z = T, y = T, x = T, v = T) + P(z = T, y = T, x = F, v = F) + \\ P(z = T, y = T, x = T, v = F) + P(z = T, y = T, x = F, v = T) \end{array}}{\begin{array}{c} P(y = T, x = T, v = T) + P(y = T, x = F, v = F) + \\ P(y = T, x = T, v = F) + P(y = T, x = F, v = T) \end{array}}$$

By using the values from NPT the following conclusion can be drawn,

$$P(z = T \mid y = T) = \frac{1 + 0 + 0 + 0}{1 + 0 + 0 + 0} = 1$$

The result states that the optional feature z is selected whenever feature y is selected. As feature y depends on feature x which is a mandatory feature, y is actually a false optional feature and hence z is also a false optional feature which will always be part of a valid product even after it is declared as an optional.

False Optional, Rule 3: A feature becomes false optional when its ancestor is a false optional feature. In the 3rd rule of false optional feature in Table 2, x is a mandatory feature and y is an optional feature of variation point v. Furthermore, z is an optional feature of y. As y is required by x, it becomes a false optional feature resulting z into a false optional feature as well. In the BN, y depends on both x and v and z only depends on y.

v	
T	F
1	0

x		
v	T	F
T	1	0
F	0	1

y			
v	x	T	F
F	F	0	1
F	T	I	I
T	F	I	I
T	T	1	0

z		
y	T	F
T	1	0
F	0	1

Fig. 4. NPT for false optional rule 3

$$P(z,y) = P(z \mid y)P(y \mid x,v)P(x \mid v)P(v)$$

$$P(z=T \mid y=T) = \frac{P(z=T,y=T)}{P(y=T)} = \frac{\sum_{x,v\in\{T,F\}} P(z=T,y=T,x,v)}{\sum_{x,v\in\{T,F\}} P(y=T,x,v)}$$

$$= \frac{\begin{array}{c} P(z=T,y=T,x=T,v=T) + P(z=T,y=T,x=F,v=F) + \\ P(z=T,y=T,x=T,v=F) + P(z=T,y=T,x=F,v=T) \end{array}}{\begin{array}{c} P(y=T,x=T,v=T) + P(y=T,x=F,v=F) + \\ P(y=T,x=T,v=F) + P(y=T,x=F,v=T) \end{array}}$$

By using the values from NPT in Fig. 4 the following conclusion can be drawn,

$$P(z=T \mid y=T) = \frac{1+0+0+0}{1+0+0+0} = 1$$

The result shows that z is a false optional feature.

Dead Feature, Rule 1: An optional feature becomes dead feature when it is excluded by a mandatory feature. In first rule for dead features in Table 2 the graph x is a mandatory feature which excludes optional feature y. Variant y can never be part of a valid configuration and thus its a dead feature. In the BN, the probability of both of these feature being selected are directly dependent on v. The probability of y being selected is also dependent on x.

v	
T	F
1	0

x		
v	T	F
T	1	0
F	0	1

y			
v	x	T	F
F	F	0	1
F	T	I	I
T	F	I	I
T	T	0	1

Fig. 5. NPT for dead feature rule 1

From the NPT in Fig. 5 the probability of selecting the variable y can be calculated by using Bayesian rule,

$$P(y,x,v) = P(y \mid x,v)P(x \mid v)P(v)$$

$$P(y=T \mid x=T) = \frac{P(x=T,y=T)}{P(x=T)} = \frac{\sum_{v\in\{T,F\}} P(x=T,y=T,v)}{\sum_{v\in\{T,F\}} P(x=T,v)}$$

$$= \frac{P(y=T,x=T,v=T) + P(y=T,x=T,v=F)}{P(x=T,v=T) + P(x=T,v=F)}$$

The probabilities of variables can be calculated from NPT as shown for other rules. Finally we get the following conclusion,

$$P(y=T \mid x=T) = \frac{0+0}{1+0} = 0$$

Thus, y becomes a dead feature.

Dead Feature, Rule 2: An optional feature becomes dead when it is excluded by a false optional feature. In the 2nd rule for dead feature in Table 2, mandatory feature x requires an optional feature y. Thus y becomes a false optional feature. Moreover, y excludes the optional feature z which implies that z is a dead feature. In the BN all of the features x, y and z are directly dependent of v. Besides, y has a dependency on x and z has a dependency on y.

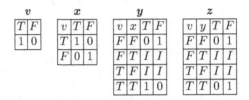

Fig. 6. NPT for dead feature rule 2

From the Fig. 6 the probability of selecting the variable z can be calculated by using Bayesian rule,

$$P(z, y, v) = P(z \mid y, v) P(y \mid x, v) P(x \mid v) P(v)$$

$$P(z = T \mid y = T) = \frac{P(z = T, y = T)}{P(y = T)} = \frac{\sum_{x, v \in \{T, F\}} P(z = T, y = T, x, v)}{\sum_{x, v \in \{T, F\}} P(y = T, x, v)}$$

$$= \frac{\begin{array}{c} P(z = T, y = T, x = T, v = T) + P(z = T, y = T, x = F, v = F) + \\ P(z = T, y = T, x = T, v = F) + P(z = T, y = T, x = F, v = T) \end{array}}{\begin{array}{c} P(y = T, x = T, v = T) + P(y = T, x = F, v = F) + \\ P(y = T, x = T, v = F) + P(y = T, x = F, v = T) \end{array}}$$

By using the values from NPT the following conclusion can be drawn,

$$P(z = T \mid y = T) = \frac{0 + 0 + 0 + 0}{1 + 0 + 0 + 0} = 0$$

The result states that the optional feature z is not selected whenever feature y is selected. y is a false optional feature and z is a dead feature as it will never be part of a valid product even after it is declared as an optional.

Dead Feature, Rule 3: A feature becomes a dead feature when its ancestor is a dead feature. In the dead feature rule 3 in Table 2, x is a mandatory feature of variation point v and y is an optional feature of v. Furthermore, y itself is a variation point and z is its optional feature. As y is excluded by x, it becomes a dead feature resulting z into a dead feature as well. In the BN x and y are dependent of v and z is only dependent of y.

v	
T	F
1	0

x		
v	T	F
T	1	0
F	0	1

y			
v	x	T	F
F	F	0	1
F	T	I	I
T	F	I	I
T	T	0	1

z		
y	T	F
T	1	0
F	0	1

Fig. 7. NPT for dead feature rule 3

$$P(z,y) = P(z \mid y)P(y \mid x,v)P(x \mid v)P(v)$$

$$P(z = T \mid y = F) = \frac{P(z = T, y = F)}{P(y = F)} = \frac{\sum_{x,v \in \{T,F\}} P(z = T, y = F, x, v)}{\sum_{x,v \in \{T,F\}} P(y = F, x, v)}$$

$$= \frac{\begin{array}{c}P(z = T, y = F, x = T, v = T) + P(z = T, y = F, x = F, v = F)+ \\ P(z = T, y = F, x = T, v = F) + P(z = T, y = F, x = F, v = T)\end{array}}{\begin{array}{c}P(y = F, x = T, v = T) + P(y = F, x = F, v = F)+ \\ P(y = F, x = T, v = F) + P(y = F, x = F, v = T)\end{array}}$$

By using the values from Fig. 7 the following conclusion can be drawn,

$$P(z = T \mid y = T) = \frac{0 + 0 + 0 + 0}{0 + 0 + 0 + 0} = 0$$

The result shows that z is a dead feature.

Dead Features, Rule 4: A feature becomes dead when it requires another dead feature. In the 4th rule of dead feature in Table 2 v is a variation point of three variant x, y and w where x is a mandatory feature and the other two are optional. Feature x and y are mutually exclusive. Thus y becomes a dead feature. Moreover, y is the variation point for another optional feature z. Since y is a dead feature, z is also a dead feature. On the contrary, the optional feature w requires z which is already a dead feature. Hence w is also a dead feature.

v	
T	F
1	0

x		
v	T	F
T	1	0
F	0	1

y			
v	x	T	F
F	F	0	1
F	T	I	I
T	F	I	I
T	T	0	1

z		
y	T	F
T	1	0
F	0	1

w			
v	z	T	F
F	F	0	1
F	T	I	I
T	F	I	I
T	T	1	0

Fig. 8. NPT for dead feature rule 4

$$P(w,y,z) = P(w \mid z,v)P(z \mid y)P(y \mid x,v)P(x \mid v)P(v)$$

$$P(w = T \mid z = F) = \frac{P(w = T, z = F)}{P(z = F)} = \frac{\sum_{x,y,v \in \{T,F\}} P(w = T, z = F, x, y, v)}{\sum_{x,yv \in \{T,F\}} P(z = F, x, y, v)}$$

All the above probabilities are calculated from Fig. 8. Finally, we derive the following conclusion which shows that z is a dead feature.

$$P(w = T \mid z = F) = \frac{0 + 0 + 0 + 0}{0 + 0 + 0 + 0} = 0$$

5   Conclusion

This paper presented an approach to define and verify software product line feature analysis rules by using Artificial Intelligence-based technique. During any software development project, handling of requirement is inherently difficult and uncertain, which make it amenable for the application of AI technique. In this work our focus was in a particular area of SE, Software Product Line. We defined a set of rules for both dead and false optional features of SPL feature diagram. BN is used to model and verify the analysis rules. In comparison to our earlier work on applying logic [17] and semantic web-based approach [18] to verify analysis rules, BN-based analysis not only manage to solve uncertain situation but also establishes a synergy between AI techniques and SPL feature analysis. Our future plan includes the extension of the analysis rules for various other analysis operations. We are also interested to apply any BN tool to perform the verification mechanically to avoid human error.

References

1. Barry, P.S., Laskey, K.B.: An application of uncertain reasoning to requirements engineering. In: Proceedings of the Fifteenth Conference on Uncertainty in Artificial Intelligence, UAI 1999, pp. 41–48. Morgan Kaufmann Publishers Inc., San Francisco (1999)
2. Benavides, D., Segura, S., Cortés, A.R.: Automated analysis of feature models 20 years later: A literature review. Inf. Syst. 35(6), 615–636 (2010)
3. Bosch, J.: Design and use of software architectures - adopting and evolving a product-line approach. Addison-Wesley (2000)
4. Clements, P.C., Northrop, L.: Software Product Lines: Practices and Patterns. SEI Series in Software Engineering. Addison-Wesley (August 2001)
5. Czarnecki, K., She, S., Wasowski, A.: Sample spaces and feature models: There and back again. In: International Software Product Line Conference, pp. 22–31 (2008)
6. Fenton, N.E., Neil, M., Marsh, W., Hearty, P., Marquez, D., Krause, P., Mishra, R.: Predicting software defects in varying development lifecycles using bayesian nets. Information & Software Technology 49(1), 32–43 (2007)
7. Jensen, F.V.: Bayesian Networks and Decision Graphs. Springer-Verlag New York, Inc., Secaucus (2001)
8. Jensen, F., Nielsen, T.: Bayesian Networks and Decision Graphs, 2nd edn. Springer, New York (2007)
9. Kang, K.C., Cohen, S.G., Hess, J.A., Novak, W.E., Peterson, A.S.: Feature-oriented domain analysis (foda) feasibility study. Tech. rep., Carnegie-Mellon University Software Engineering Institute (November 1990)

10. von der Massen, T., Lichter, H.: Deficiencies in Feature Models. In: Mannisto, T., Bosch, J. (eds.) Workshop on Software Variability Management for Product Derivation - Towards Tool Support (2004)
11. de Melo, A.C.V., de J. Sanchez, A.: Software maintenance project delays prediction using bayesian networks. Expert Syst. Appl. 34(2), 908–919 (2008)
12. Meziane, F., Vadera, S.: Artificial intelligence applications for improved software engineering development: New prospects. IGI Global, Hershey (2009)
13. Pearl, J.: Probabilistic Reasoning in Intelligent Systems: Networks of Plausible Inference. Morgan Kaufmann Publishers Inc., San Francisco (1988)
14. Pendharkar, P.C., Subramanian, G.H., Rodger, J.A.: A probabilistic model for predicting software development effort. IEEE Trans. Software Eng. 31(7), 615–624 (2005)
15. Radliński, L., Fenton, N., Neil, M., Marquez, D.: Improved decision-making for software managers using bayesian networks. In: Proceedings of the 11th IASTED International Conference on Software Engineering and Applications, SEA 2007, pp. 13–19. ACTA Press, Anaheim (2007)
16. Rincón, L.F., Giraldo, G., Mazo, R., Salinesi, C.: An ontological rule-based approach for analyzing dead and false optional features in feature models. Electr. Notes Theor. Comput. Sci. 302, 111–132 (2014)
17. Ripon, S., Azad, K., Hossain, S.J., Hassan, M.: Modeling and analysis of product-line variants. In: de Almeida, E.S., Schwanninger, C., Benavides, D. (eds.) SPLC (2), pp. 26–31. ACM (2012)
18. Ripon, S., Piash, M.M., Hossain, S.A., Uddin, S.: Semantic web based analysis of product line variant model. International Journal of Computer and Electrical Engineering (IJCEE) 6(1), 1–6 (2014)
19. del Sagrado, J., del Águila, I.M., Orellana, F.J.: Architecture for the use of synergies between knowledge engineering and requirements engineering. In: Lozano, J.A., Gámez, J.A., Moreno, J.A. (eds.) CAEPIA 2011. LNCS (LNAI), vol. 7023, pp. 213–222. Springer, Heidelberg (2011)

A Content-Based Approach for User Profile Modeling and Matching on Social Networks

Thanh Van Le, Trong Nghia Truong, and Tran Vu Pham

Faculty of Computer Science and Engineering
Ho Chi Minh City University of Technology, VNU-HCM
No 268, Ly Thuong Kiet Street, District 10, Ho Chi Minh City, Vietnam
{ltvan,ttnghia,t.v.pham}@cse.hcmut.edu.vn

Abstract. The development of social networks gives billions of users the convenience and the ability to quickly connect and interact with others for raising opinions, sharing news, photos, etc. On the road for building tools to extend friend circles as large as possible, one of the most important functions of a social network is the recommendation which proposes a group of people having some common characteristics or relations. A majority of social networks have friend suggestion function based on mutual friends. However, this suggestion mechanism does not care much about the actual interests of the users hidden in his comments, posts or activities. This paper aims to propose a profile modeling and matching approach based on Latent Dirichlet Allocation (LDA) and pretopological-based multi-criteria aggregation to explore topics that exist in user posts on a social network. We explored interesting points of pretopology concepts - a mathematical tool - and applied them for better solving the raised problem. This approach allows us to find out users who have similar interests and also other information involving user profiles.

Keywords: LDA, hidden topics, profile matching, social networks, pretopology.

1 Introduction

User profile comparison is not a new research area. There have been a lot of efforts from research community spent on the area of user profiling and matching. A visible example is follower suggestion in Twitter. A user usually follows events or other users that they are interested in [1]. In Facebook, connection links (mutual friends) are used for friend suggestion in order to broaden the relationship of a user. With retailers, user similarity is an important factor. For instance, similar user profiles that satisfy a standard set can be grouped together so that they can predict future behaviors of each user based on knowledge of the group they belong to [2].

One of popular approaches for suggesting goods to potential buyers in retailer Web sites is collaborative filtering [3]. It is widely used to recommend users to other users or items to users. Its effectiveness has been proved by many experiments in [4], [5] with an assumption that those users having similar profile will share same tastes about what they like, buy or rate.

In addition to collaborative filtering, there also exist many methods based on content filtering [6], [7]. These methods rely on item's description or content that has been rated

M.N. Murty et al. (Eds.): MIWAI 2014, LNAI 8875, pp. 232–243, 2014.

and bought such as the title of a film that users watched and rated, content of a book which they read. Content filtering methods can also be applied to digital content of music by analyzing its rhythm or harmony [7]. With text-based content, there are many ways to profile and compare user interests. One can vectorize the text to build profiles, then compare them by using cosine similarity or pearson correlation. Some advanced methods such as Latent Semantic Analysis (LSA), probabilistic Latent Semantic Analysis (pLSA) or LDA can also be used to reduce the dimension of text to get more precise results [8].

Another content-based filtering approach integrated in Apache Lucene was proposed in [6]. In this method, a collection of document represented for a user is tokenized and indexed to a large collection of text (terms) by all users. To detect neighbors of one target user, a query is constructed by using the target's text. This text will also be tokenized, indexed, and compared with indexed terms of other users to compute the similarity between target's terms and other users' terms. The result of term comparison will be ranked. The group of users that have the highest ranks form the neighborhood of the target user.

Moreover, using ontology is also a popular approach in Information Retrieval [9]. Topics and their relationships can be organized in a hierarchical structure using ontology. A user profile can then be presented by a set of pre-defined topics in the ontology. Each topic has a interest score that measures how much it is cited in this ontology. Then the set of topics is turned into a flat vector based on ontology, and the matching between two users is calculated as the similarity between two vectors.

In a social network, profile matching between two users can be done by comparing their connections on the network. A connection of a user can be defined as a relationship between the user and his/her friends, comments, favourites, and shared items. That information will be presented as a graph in which the users and other entities are nodes and the interactions between them are labeled as edges. Then the profile matching between two users is turned into analyzing the similarity of two nodes representing each of them.

In this paper we propose a content-based approach using LDA for exploring hidden topics in user posts and some other social features. In addition to relatively static features that are available on most social networks, for example, birthday, hometown and place, the most important information of a user profile that we are considering is the user's interest. In a social network, a user's interest is usually reflected his/her posts such as comments, feeds, photos, etc. For text-based features such as feeds and comments, LDA is used as tool for extracting topics. Then a suitable similarity measure will be applied to find out the neighborhood of a certain user. Some experiments with data crawled from a Vietnamese social networks are also reported in this paper.

This paper is divided into five sections including this one. The second section recalls preliminaries about LDA model and Vietnamese language processing. Our proposed approach and some experimental results are presented in details in next two sections. The last section concludes all steps of our works.

2   Preliminaries

In this section, we are going through some basic theories to be used in subsequent sections. Pretopological space which is the core background of our work would be first presented. Then we introduce LDA topic model along with Gibbs sampling to generate topic model, and previous works on word segmentation into tokens for Vietnamese language.

2.1   Pretopology Concepts

Pretopology was first introduced in [10] by a research group called Z.Belmant. Based on topology concepts and its characteristics, they proposed a generalization of topology space but more abstract called pretopology - a mathematical tool which defines proximity phenomenon based on two principal functions of adherence (also called pseudoclosure) $a(.)$ and interior $i(.)$ In this paper we focus on only the function of pseudoclosure.

Definition 1. *Consider a set E and $P(E)$ the set of subsets of E. The pseudoclosure $a(.)$ is defined as follows:*

– $a(\emptyset) = \emptyset, \forall A \in P(E), A \subset a(A)$

A set E with the pseudoclosure $a(.)$ is called pretopological space, denoted (E,a). Contrary to what happens in classical topology, this function is not necessarily idempotent. Therefore, the pseudoclosure can be used to express the extension process as toxic/pollution expansion, patient pathology clustering by pretopologogy neighbor systems, etc. Given remarks on the non idempotence of the pseudoclosure function, we can model a process as below:

$$A \subset a(A) \subset a(a(A)) \subset ... \subset a^n(A) \subset ... \subset E$$

Let (E,a) a pretopological space. According to the properties of the function pseudoclosure, we obtain different types of pretopological space. The most interested type in our work is the space of type V.

Definition 2. *The space (E,a) is the pretopological space of type V if it verifies the following properties:*

$$\forall A \in P(E), \quad \forall B \in P(E) \quad : \quad A \subset B \Longrightarrow a(A) \subset a(B)$$

Some important properties of type V allow us to derive a notion of closure $F(.)$ based on the pseudoclosure function of a finite set E:

$$\exists k < |E|, A \in P(E), F(A) = a^k(A)$$

Moreover, the pseudoclosure of type V can be produced based on a binary relationship family. A dataset in which elements are connected by one or more binary relations R_i at least reflexive, we can consider $B_i(x)$ defined as follows:

$$B_i(x) = \{y \in E | xR_iy\}$$

Let $V(x)$ the pre-filter of $P(E)$, it defines the neighborhoods family of x for a pretopological space of type V.

$$\forall A \in P(E), a(A) = \{x \in E | \forall V \in V(x), V \cap A \neq \emptyset\}$$

The above properties is too important because we will use it to integrate multi-criteria for proximity measure of neighborhood. As demontrated in [11], we could also produce data clusters if our dataset space exists a symmetric distance. We will apply this approach for solving our problem in finding similar user profiles of a Vietnamese social network.

2.2 LDA Topic Model

LDA is an generative model - a model can generate observable data, first introduced by Blei et al. in 2003 [12]. The main idea of this model is that a document is built up from a list of fixed topics with probability over each topic, meanwhile a topic itself is also built up from a fixed number of words with different probability. The model can be presented by the following plate notation as shown in Fig. 1.

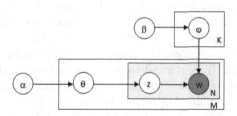

Fig. 1. LDA's plate notation

- α, β are Dirichlet priors.
- θ_i, φ_k are the topic distribution over document i and word distribution over topic k.
- z_{ij} is a topic of a word in a document.
- w_{ij} is a specific word in a document.
- K is the number of topics.
- M is the number of documents.
- N is the number of words in a document.

To create a document i with this model, we first define a number of words, then choose two variables $\theta_i \sim Dir(\alpha)$ and $\varphi_k \sim Dir(\beta)$ where $Dir(\alpha), Dir(\beta)$ are the Dirichlet distribution with parameters. After that, we continue choosing each word in the document by choosing a topic $z_{ij} \sim Mul(\theta_i)$ with a word $w_{ij} \sim Mul(\varphi_{(z_i j)})$ where $Mul(\theta_i)$ and $Mul(\varphi_{(z_i j)})$ are the Multinomial distribution with corresponding parameters.

2.3 Collapsed Gibb Sampling

The Collapsed Gibbs Sampling (CGS) [13] is a process used to discover topic model that has generated documents within a corpus. In original paper, Blei et al. also proposed a method which serves the same purpose called variational Bayersian inference but CGS uses lower operation to attain the same perplexity while evaluating [13], [14]. It uses the Markov Chain Monte Carlo to calculate posterior probability φ and θ for the model by looping through a number of documents and words, assigning required parameters, then calculating final posterior. In CGS method, the θ and φ are omitted instead of directly sampling, only z is sampled, then two distributions can be derived from z by a simple formula.

2.4 Vietnamese Text Processing

Processing Vietnamese text is not the same as processing English text. Segmenting text in English usually includes removing stop words, stemming, and identifying word boundary. Word boundary can be detected by using special characters such as period, space, and quote or the capital letter at the first position of a word. However, in Vietnamese language, there is a large number of compound words which are constructed by more than one syllable. Syllables in a compound word are also separated by space(s). This leads to ambiguity for an algorithm to detect whether a syllable in a sentence is a single word or belongs to a compound word. To segment Vietnamese document, there are two popular programs known as vnTokenizer and JVnSegmenter. vnTokenizer is more stable than JVnSegmenter [15] and can gain accuracy over 90% [16]. With vnTokenizer, an input text is analyzed to detect common patterns such as name, number, date, email address, etc. If input text does not occur as common patterns but appears as a phrase, a representation graph is built and maximal matching algorithm is used to detect valid combination within phrases and groups of ambiguous words. These ambiguous groups are resolved by a bigram language model which is trained using a manual segmentation text dataset.

3 Content-Based Profiling and Matching

Before introducing our proposed approach, we start with examining which features in a social network can affect the similarity calculation between two users. In a general social network, the basic information about a user are hobbies, working history, education, hometown, friends and ever changing of his/her activities, interactions with friends and with the community. To compute the similarity or dissimilarity between two users, we have to transform the information into a standard form and have a suitable measure to compare information of a certain data type. Firstly, we show the process of building each user profile by getting the most possible attributes on a social network. Then we suggest a way to filter and compare users with each other.

3.1 Building User Profiles

We use similar profiling method presented in [6]. Each user can be defined by a set of following general attributes:

- Single-valued attribute: the attribute has only one value such as gender, religion, name.
- Multiple-valued attribute: the attribute has two or more values such as languages, jobs, universities.
- Hierarchical attribute: the attribute splits into different levels such as address, interest.
- Content attribute: the attribute include text from user activities such as posting a status, commenting on a post.

With each attribute, a proper method to compare them is required. The next section deals with matching attributes of different users.

3.2 Matching Strategy

Finding users having similar profiles to a given user profile is a multi-step process. We should first create a candidate list of users who are potentially similar to the given one to reduce the search space. Then for each user in the short list, we use two different strategies to generate a list of relevant users to the given one. Both strategies involve comparing attributes of two users. The method used to compare attributes of two different users is detailed below.

Building Candidate List. A candidate list for a certain user can be constructed based on the mutual friend intuition that says: "if many of my friends consider Alice a friend, perhaps Alice could be my friend too" [17]. It is better that we do not compare a user with all others in a social network because of its complexity and heterogeneous. For a user x, its recommendation candidate set is defined as:

$$RC(x) = \{\forall u, y \in F(x) | F(u,y)\}$$

where:

- $RC(x)$ is candidate set of user x.
- $F(x)$ is the friend set of user x.
- $F(u,y)$ denotes that u and y are friends.

The candidate set is ranked by the size of their mutual friends. Therefore, if the two users have more common friends they will be more likely to be friends with each other. Our simple calculation shows that about 75% of the friends is in this candidate list formula. If we go deep further by one level on friend, the number of users is 96% but the search space increases more than ten times. This will cause a significant slowdown on speed of processing. Therefore, we decide to stop at one level friend.

Single-Value Attribute Matching. The single-value attribute of a user x can be presented by a set with only one element: $A_{single}(x) = \{a_x\}$. When matching these attributes, the following equation is used:

$$f_{single}(x,y) = \begin{cases} 1 & \text{when } A_{single}(x) = A_{single}(y) \\ 0 & \text{when } A_{single}(x) \cap A_{single}(y) = \emptyset \end{cases} \tag{1}$$

f_{single} is the matching score when comparing single-value attributes of two users x and y.

Multi-value Attribute Matching. Multi-value attribute of a user x is presented by a set with variable number of elements: $A_{multi}(x) = a_1, a_2, a_3, ..., a_n, n \in N$. The matching score between two sets of attributes of any two users is calculated using the following formula:

$$f_{multi}(x,y) = \frac{|A_{multi}(x) \cap A_{multi}(y)|}{M} \tag{2}$$

where:

- $f_{multi}(x,y)$ is the matching score of multi-value attribute between two users x and y.
- $|A_{multi}(x) \cap A_{multi}(y)|$ is the number of elements of intersection set between users x and y.
- $M = |A_{multi}(x)|$ if $|A_{multi}(x)| \geq |A_{multi}(y)|$, otherwise $M = |A_{multi}(y)|$

Hierarchical Attribute Matching. The hierarchical attribute of a user x is presented by an ordered set. Each element of the set contains a level with its score: $A_{hier}(x) = \{\{lv_1, s_1\}, \{lv_2, s_2\}, ..., \{lv_n, s_n\}\}$. The matching score between two attributes can be calculated as:

$$f_{hier}(x,y) = \begin{cases} max(s_i, s_j) \ \forall j \neq i, \{lv_i, s_i\}, lv_j, s_j \subset (A_{hier}(x) \cap A_{hier}(y)) \\ 0 \qquad\qquad\qquad\qquad\qquad\qquad A_{hier}(x) \cap A_{hier}(y) = \emptyset \end{cases} \tag{3}$$

where:

- $f_{hier}(x,y)$ is the matching score of hierarchical attribute between two users x and y.
- $max(s_i, s_j)$ is the function that returns the biggest value between s_i and s_j.

Continuous Attribute Matching. The continuous attribute is presented as: $A_{cont}(x) = \{c_x\}, c_x \in R$. To compare those types of attributes, we define a set of range $G = \{g_1, g_2, g_3, ..., g_n\}$ and a function $f(x) : A_{cont}(x) \rightarrow G$. Then, the matching score is calculated as:

$$f_{cont}(x,y) = 1 - \frac{|f(A_{cont}(x)) - f(A_{cont}(x))|}{M} \tag{4}$$

$f_{cont}(x,y)$ is the matching score of continuous attribute between two users x and y.

Content Attribute Matching. For a given profile, the obtained result after creating LDA vector process is a vector with k elements. Each element is the probability distribution over a corresponding topic. This attribute of a user x can be presented: $A_{content}(x) = \{x_0, x_1, x_2, ..., x_{(k-1)}\}$. To calculate matching score between these two vectors (also called similarity coefficient), we use widely accepted cosine method:

$$f_{content}(x,y) = \frac{\sum\limits_{i=0}^{k-1} x_i y_i}{\sum\limits_{i=0}^{k-1} x_i^2 . \sum\limits_{i=0}^{k-1} y_i^2} \tag{5}$$

$f_{content}(x,y)$ is the matching score of content attribute between two user x and y.

3.3 Generating Subsets of Similar Users

In this section we propose two methods for generating the similar user set of a given user x after creating the candidate list. One method is based on total similarity between x and other users in candidate list via choosing the best combination of weights of each comparable attribute. Another method is based on pretopology to take advantages of the relations between users in a network.

Generate Similar User Set by Finding the Best Weights Combination of Partial Score (BW) Firstly, we define the total score is the weighted sum of all partial scores computed by equations (1) - (5):

$$f_{total}(x,y) = \omega_1 f_1(x,y) + \omega_2 f_2(x,y) + \omega_3 f_3(x,y) + ... + \omega_n f_n(x,y)$$

where:

- ω_i is the weight of each attribute.
- $f_1(x,y), f_2(x,y), f_3(x,y), .., f_n(x,y)$ are the scores of each attribute when comparing.
- $\omega_1 + \omega_2 + \omega_3 + ... + \omega_n = 1$

Assume the candidate of a given user x:

$$RC(x) = \{m_i\}, i \in N.$$

Then we have to select the best weight combination and define a positive natural number K to get the top K users in $RC(x)$ that have the biggest total score in comparison with user x.

Generate Similar User Based on Pretopology (PB). With this method, we define a set of thresholds $\{\varepsilon_i\}, i \in N$ corresponding to each attribute to be compared and a natural number γ less than total number of attributes. We define a function $R(x)$ that describes the neighborhood of a user x as follows:

$$R(x) = \{\{m_i, s_i\}, i \in N, s_i \geq e_i\}$$

This formula means that if a user has a certain number of scores pass its threshold, then it is considered as a neighborhood of user x. We define a pseudo closure $a(.)$ for a set A as follows:

$$a(A) = \{x \in E, R(x) \bigcap A \neq \emptyset\}$$

Then we create a pretopology space type V, which is the subset of E with the pseudo closure $a(.)$. The initial set A contains only one element x. An element existing in $a(x)$ means that it has a strong relation with x. The closure $F(x)$ forms a cluster of users. When finding similar user set by PB method, it is not necessary to determine weighted scores for each attribute as BW method does. In this method, we care much about relationships and the expansion connection between users based on characteristics of pseudoclosure fucntion. Thus this method is considered as a suitable idea to enlarge friend circles or measure the spreading affectation of a user.

4   Experiment Results

Dataset used in our experiments was crawled from "Truong Xua" [1], a social network that attracts users by bringing back their memories about previous schools, classes they were in. Most users in this network use their real names, school names and classes they studied to find their old friends and make new friends as well as update their status, publish blogs and interact with others via comments, likes and chat.

Total number of users of this social network was about 2.5 millions at May, 2013. We could select random users in this network to create the dataset but many of them are inactive. Therefore, we began with a sample of 40 active seed profiles from the website, and then we followed friend connections of each user to get more users. Finally, the total number of user profiles crawled were about 507000. For each user, the following information was taken into account:

- Personal information: place (of living), hometown, gender, birthday, marital status, school name with class and years .
- Activity: blog, share, like, status, chat log. In blog and status there are comments and likes, but we can only get all comments that contain commenter identifier and its content. Likes and shares are listed as top 10 last actions so we do not care about them at all.
- Friend list: we get all ids of friends of a user

We did a simple calculation to get some basic information about the dataset as shown in Table 1:

Table 1. Information about experimental dataset

Average friend per user	24
Total number of activity	498328
Number of users that have no activity	388862
Number of users that have local information	74760
Number of users that have birthday information	489251
Number of users that have school information	219812

The raw text crawled from the website was processed to eliminate stop-words, create a list of tokens and its corresponding frequency for further tasks of word filtering. From the token list, LDA model was built without removing rare words and common words. The result was a set of topics with no clear difference between them because of the domination of high frequency words. Words that had frequency in range 20-20000 were kept separately to create LDA model with these parameters: $\alpha = 1.0, \beta = 0.01, T = 200$ (T was the number of topics to be created). These values of parameters were chosen as they produced the best result in topic detection by LDA model.

In this specific dataset, for each type of attribute a criterion was chosen. There were five attributes in this dataset. Each of them had an appropriate similarity measure:

---

[1] www.truongxua.vn

- School information shared the same matching method with multi-value attribute because a user usually had a number of different schools.
- Living place and hometown had hierarchical structure with three levels, each level was associated with a score: ward, 1.0, district, 0.75, national, 0.
- Age was mapped into sets of range 10-15, 15-20, 20-25, ..., 55-60 and we calculated similarity between two user by using equation (4).
- Blog and status were sampled with the LDA model above to create a vector having 200 dimension, then cosine similarity was used to indicate score between two users.

Then we examined two different methods to generate the similar user sets with different parameters. First, we detected the best weight combination of BW method by selecting 46 different combinations of five partial score and observed how many of removed friends could be recovered in top 50 users in its similar set. The best weight combination was: $\omega_{school} = 0.5, \omega_{place} = \omega_{hometown} = \omega_{age} = \omega_{content} = 0.125$.

We also chosen three threshold sets for PB method. Each set was presented as: $\{\varepsilon_{school}, \varepsilon_{place}, \varepsilon_{hometown}, \varepsilon_{age}, \varepsilon_{content}\}$. Then we chose the parameters such that the average output size of both methods was nearly equivalent. The result of testing for retrieved friends is shown in the following table:

Table 2. Friend circles inferred from BW and PB

BW Method		PB Method		
K	Number of friends	Threshold set	Pseudoclosure level	Number of friends
80	32901	{0.3, 0.75, 0.75, 0.5, 0.3}	1	23521
64	31349	{0.5, 0.75, 0.75, 0.7, 0.5}	1	19771
3	5138	{0.7, 1.0, 1.0, 1.0, 0.7}	1	5621
7	10490		2	9367
15	18001		3	10240
27	24201		4	10508

We can observe that the number of retrieved friends recommended by BW method is larger than those by computed by PB method. To evaluate the correctness of users recommended by PB and BW methods, we extracted randomly 400 users and computed theirs circles. Precision at k (*Precision@k*) was used to compute the percentage of relevant users in top K retrieved users from PB and BW methods.

Table 3. Precision at 20, 30, 40 of BW and PB

Method	Precision@20	Precision@30	Precision@40
BW	0.9658	0.7970	0.6149
PB	0.9629	0.8039	0.6337

Table 4 shows that the precision at 20, 30, 40 users of two methods are quite similar. We continued computing average similar coefficient for each method and realized that

PB gave the better result. Fig. 2 illustrates distribution of similar coefficient. A weakness of BW method that can be easily observed is that it always returns suggested friends although they are too different (similar coefficient close to 0). In contrast, PB method may lead to null result as it does not find any user matching with the criteria defined for the neighborhood family. However, this is considered as one of benefits of this method as it proposes only users that have a strong connection.

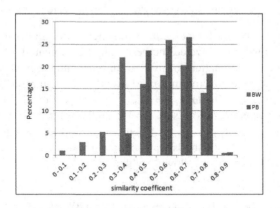

Fig. 2. Distribution of similar coefficient

5   Conclusion

As reported in this paper, our proposed method shows that a user in a social network can be profiled by activities, along with personal information such as age, place, hometown, list of friends, and interactions with others. The text-based features are segmented to build a token set. The token set are then continually filtered by term-frequency to get only significant tokens which contribute well to the topic modeling. Then a topic model is built for exploring hidden topics from text of users. In order to avoid verifying all users for finding the neighborhood of a target user, social features should be used for generating a candidate list for comparing users with high score without running over the entire network. To suggest new friends for a given user, BW and PB methods are considered as good solutions. BW method compute more faster and always return nearest friends, but sometime it could get a new friend who is totally different because this user is the "nearest" neighbor. PB method focuses much on relationship between users and only returns a new friend for a specific user if they truly have connection in their profiles.

Acknowledgement. This research is partially funded by Vietnam National University Ho Chi Minh City (VNU-HCM) under grant number B2014-20-07.

References

1. Guy, I., Ronen, I., Wilcox, E.: Do you know?: recommending people to invite into your social network. In: Proceedings of the 14th International Conference on Intelligent User Interfaces, pp. 77–86. ACM, NY (2009)
2. Park, Y.J., Chang, K.N.: Individual and group behavior-based customer profile model for personalized product recommendation. Expert System Application 36(2), 1932–1939 (2009)
3. Breese, J.S., Heckerman, D., Kadie, C.: Empirical analysis of predictive algorithms for collaborative filtering. In: Proceedings of the Fourteenth Conference on Uncertainty in Artificial Intelligence, pp. 43–52. Morgan Kaufmann Publishers Inc., CA (1998)
4. Linden, G., Smith, B., York, J.: Amazon.com recommendations: Item-to-item collaborative filtering. IEEE Internet Computing 7(1), 76–80 (2003)
5. Li, Y., Lu, L., Xuefeng, L.: A hybrid collaborative filtering method for multiple-interests and multiple-content recommendation in e-commerce. Expert Systems with Applications 28, 67–77 (2005)
6. Hannon, J., Bennett, M., Smyth, B.: Recommending twitter users to follow using content and collaborative filtering approaches. In: Proceedings of the Fourth ACM Conference on Recommender Systems, pp. 199–206. ACM, NY (2010)
7. McFee, B., Barrington, L., Lanckriet, G.R.G.: Learning content similarity for music recommendation. IEEE Transactions on Audio, Speech and Language Processing, 2207–2218 (2012)
8. Kakkonen, T., Myller, N., Sutinen, E., Timonen, J.: Comparison of dimension reduction methods for automated essay grading. Educational Technology and Society 11(3), 275–288 (2008)
9. Sieg, A., Mobasher, B., Burke, R.: Improving the effectiveness of collaborative recommendation with ontology-based user profiles. In: Proceedings of the 1st International Workshop on Information Heterogeneity and Fusion in Recommender Systems, pp. 39–46. ACM, NY (2010)
10. Belmandt, Z.: Manuel de Prétopologie et ses Applications. Hermes Sciences Publications (1993)
11. Le, T.V., Truong, T.N., Nguyen, H.N., Pham, T.V.: An efficient pretopological approach for document clustering. In: 2013 5th International Conference on Intelligent Networking and Collaborative Systems (INCoS), pp. 114–120. IEEE (2013)
12. Blei, D.M., Ng, A.Y., Jordan, M.I.: Latent dirichlet allocation. Journal of Machine Learning Research 3, 993–1022 (2003)
13. Griffiths, T.L., Steyvers, M.: Finding scientific topics. Proceedings of the National academy of Sciences of the United States of America, NY, USA, PNAS, 5228–5235 (2004)
14. Asuncion, A., Welling, M., Smyth, P., Teh, Y.W.: On smoothing and inference for topic models. In: Proceedings of the Twenty-Fifth Conference on Uncertainty in Artificial Intelligence, pp. 27–34. AUAI Press, Virginia (2009)
15. Dinh, Q.T., Le, H.P., Nguyen, T.M.H., Nguyen, C.T., Rossignol, M., Vu, X.L.: Word segmentation of vietnamese texts: a comparison of approaches. In: Proceedings of 6th International Conference on Language Resources and Evaluation, Morocco (2008)
16. Pham, D.D., Tran, G.B., Pham, S.B.: A hybrid approach to vietnamese word segmentation using part of speech tags. In: Proceedings of the 2009 International Conference on Knowledge and Systems Engineering, pp. 154–161. IEEE Computer Society, Washington, DC (2009)
17. Chen, J., Geyer, W., Dugan, C., Muller, M., Guy, I.: Make new friends, but keep the old: recommending people on social networking sites. In: Proceedings of the SIGCHI Conference on Human Factors in Computing Systems, pp. 201–210. ACM, NY (2009)

Author Index